Soil Microbiology

Soil Microbiology

Dominique Salgueiro

Larsen & Keller
www.larsen-keller.com

Soil Microbiology
Dominique Salgueiro
ISBN: 978-1-64172-607-8 (Hardback)

Larsen & Keller

Published by Larsen and Keller Education,
5 Penn Plaza,
19th Floor,
New York, NY 10001, USA

Cataloging-in-Publication Data

Soil microbiology / Dominique Salgueiro.
    p. cm.
Includes bibliographical references and index.
ISBN 978-1-64172-607-8
1. Soil microbiology. 2. Microbiology. 3. Soil biology. I. Salgueiro, Dominique.
QR111 .S65 2022
578.757--dc23

For more information regarding Larsen and Keller Education and its products, please visit the publisher's website www.larsen-keller.com

Table of Contents

Preface   **VII**

Chapter 1 **Introduction to Soil Environment**   **1**
    a. Soil   1
    b. Soil Fertility   23
    c. Soil Metagenomics   25
    d. Soil as a Habitat   32
    e. Soil Biota   40

Chapter 2 **Understanding Soil Microbiology**   **44**
    a. Soil Microbes   48
    b. Compost Microorganisms   56
    c. Microbial Biomass   59
    d. Organic Matter and Microorganisms in Soil   62
    e. Role of Soil Microorganisms in Plant Mineral Nutrition   64
    f. Carbon Cycle   83
    g. Nutrient Cycling   85
    h. Phosphorus Cycle   91
    i. Factors affecting Microbial Community in Soil   92
    j. Soil Microbial Interactions and Organic Farming   95
    k. Types of Soil Microbes to Nurture Plant and Soil Health   100
    l. Soil Microbes in Plant Sulphur Nutrition   101

Chapter 3 **Bacteria and Archaea in Soil**   **109**
    a. Soil Bacteria   109
    b. Actinobacteria   114
    c. Rhizobia   121
    d. Actinomycetes   139
    e. Archaea in Dry Soil Environments   143
    f. Nitrogen Fixation   153
    g. Harmful Bacteria in Soil   155

Chapter 4 **Soil Fungus and Virus**   **160**
    a. Soil Fungus   160
    b. Common Types of Fungi Found in Soil   164
    c. Mycorrhizae   165
    d. Role of Soil Fungus   172
    e. Soil Virus   175

Chapter 5 **Soil Nematodes and Protozoa**   **199**
    a. Soil Nematodes   199
    b. Benefits of Nematodes in Healthy Soil Ecosystems   203

c. Soil Nematodes in Organic Farming Systems   205
d. Chemical Control of Nematodes   209
e. Soil Protozoa   214
f. Soil Protozoa as Bioindicators   217

Chapter 6   **Applications of Soil Microbiology**   **234**
a. Standardisation of Methods in Soil Microbiology   234
b. Soil Microorganisms and Global Climate Change   239
c. Effects of Heavy Metal Contamination upon Soil Microbes 243

Permissions

Index

Preface

The study of microorganisms which are present in the soil is referred to as soil microbiology. It focuses on the functions of microorganisms as well as the effect which they have on the properties of soil. Such microorganisms are classified into diverse categories such as bacteria, algae, fungi, protozoa and actinomycetes. Each group has specific characteristics and functions in the soil. These microbes make the soil rich in nutrients and minerals and produce hormones that encourage growth. They also help in stimulating the immune system of plants. Healthy soil increases fertility in various ways. It supplies nutrients such as nitrogen and protects against pests and diseases. This book aims to shed light on some of the unexplored aspects of modern soil microbiology. Some of the diverse topics covered herein address the varied branches that fall under this category. This textbook will serve as a valuable source of reference for those interested in this field.

Given below is the chapter wise description of the book:

Chapter 1- Soil is the combination of various materials such as minerals, liquids, organic matter, gases and organisms that together support life. This is an introductory chapter which will introduce briefly all the significant aspects of soil environment such as fertility, metagenomics and biota, along with the physical and chemical properties of soil.

Chapter 2- Soil microbiology is a domain that deals with the study of microorganisms in soil. It also studies their functions and their effect on soil properties. Microbes play a vital role in making the nutrients and minerals in the soil available to plants. This chapter has been carefully written to provide an easy understanding of the varied facets of soil microbiology as well as the different kinds of microorganisms which are found in soil.

Chapter 3- There are several categories of microbes which are present in the soil such as bacteria, actinomycetes, algae, protozoa and fungi. The most important function of bacteria and archaea is to fix nitrogen in the soil. The chapter closely examines the different types of bacteria in soil such as actinobacteria and rhizobia to provide an extensive understanding of the subject.

Chapter 4- Fungi play an important role in the soil in various capacities such as food sources for larger organisms and beneficial symbiotic relationships with plants. Viruses also play a significant role in the soil due to their ability to transfer genes from host to host. The topics elaborated in this chapter will help in gaining a better perspective about the different fungi and viruses found in the soil as well as the roles which they play.

Chapter 5- Nematodes are non-segmented worms. They are beneficial to the soil as they play a vital role in controlling disease and cycle nutrients. Protozoa are eukaryotic organisms that primarily feed on bacteria and release excess nitrogen in the process. The chapter closely examines the key functions of nematodes and protozoa in the soil as well as their use for agricultural purposes.

Chapter 6- Soil microbiology is applied in a number of different areas, such as agriculture and for the production of biopesticides. It is also involved in studying the relationship between global climate change and microbes as well as the effect of heavy metal contamination upon the microorganisms in the soil. This chapter closely examines these important applications of soil microbiology to provide an extensive understanding of the subject.

At the end, I would like to thank all those who dedicated their time and efforts for the successful completion of this book. I also wish to convey my gratitude towards my friends and family who supported me at every step.

Dominique Salgueiro

Introduction to Soil Environment

Soil is the combination of various materials such as minerals, liquids, organic matter, gases and organisms that together support life. This is an introductory chapter which will introduce briefly all the significant aspects of soil environment such as fertility, metagenomics and biota, along with the physical and chemical properties of soil.

Soil

Soil is the loose surface material consisting of inorganic particles and organic matter that covers most of the land surface. Soil provides the structural support and the source of water and nutrients for plants used in agriculture.

Soils vary greatly in their chemical and physical properties which depend on their age and on the conditions (parent material, climate, topography and vegetation) under which they were formed.

Processes such as leaching, weathering and microbial activity combine to make a whole range of different soil types, each of which has particular strengths and weaknesses for agricultural production.

Inorganic Component

Inorganic material is the major component of most soils. It consists largely of mineral particles with specific physical and chemical properties which vary depending on the parent material and conditions under which the soil was formed. It is the inorganic fraction of soils which determines soil physical properties such as texture and has a large effect on structure, density and water retention.

Soil Texture

The texture of soil is a property which is determined largely by the relative proportions of inorganic particles of different sizes.

In Australia the following five size fractions are used to describe the inorganic fraction of soils:

- Gravel: particles greater than 2 mm in diameter.

- Coarse sand: particles less than 2 mm and greater than 0.2 mm in diameter.

- Fine sand: particles between 0.2 mm and 0.02 mm in diameter.

- Silt: particles between 0.02 mm and 0.002 mm in diameter.

- Clay: particles less than 0.002 mm in diameter.

Sand

Quartz is the predominant mineral in the sand fraction of most soils. Sand particles have a relatively small surface area per unit weight, low water retention and little chemical activity compared with silt and clay.

Silt

Silt has a relatively limited surface area and little chemical activity. Soils high in silt may compact under heavy traffic and this affects the movement of air and water in the soil.

Clay

Clays have very large surface areas compared with the other inorganic fractions. As a result clays are chemically very active and are able to hold nutrients on their surfaces. These nutrients can be released into soil water from where they can be used by plants. Like nutrients, water also attaches to the surfaces of clays but this water can be hard for plants to use.

There are many different types of clays. The ability of clays to swell and to retain a shape into which they have been formed, as well as their sticky nature, distinguish them from sand and silt.

Soil Textural Class

The relative proportion of sand, silt and clay particles determines the physical properties of soil including the texture. The surface area of a given amount of soil increases significantly as the particle size decreases. Consequently, the soil textural class also gives an indication of some soil chemical properties.

The exact proportions of sand, silt and clay in a soil can only be determined in a laboratory but a naming system has been developed to approximately describe the relative proportions of sand, silt and clay in soil. This classification of soil can be undertaken in the field where particular properties indicate possible textural classes.

To estimate texture in the field, crush a small sample of soil (10 to 20g) in one hand. After removing any gravel or root matter, work the soil in the fingers to break down any aggregates which may be present. With the sample moist but not sticky, the textural class can be estimated by the feel of the sample between the fingers.

Textural Class Description

Sand: A sand has a loose gritty feel and does not stick together. Individual sand grains can be seen or felt.

Loamy Sand: In a loamy sand particles barely stick together and a moulded piece of soil just holds its shape.

Sandy Loam: A sandy loam sticks together more than a loamy sand but can be easily broken. Individual sand grains can be felt and heard if a wet sample is rubbed between the index finger and thumb and held close to the ear.

Silty Loam: A silty loam is like a loam but has a smooth silky feel when a moist sample is pushed between the index finger and thumb. On drying a sample can form a hard lump but this may be broken by hand.

Loam: A loam breaks into crumbs but will tend to stick together. Sand grains cannot be felt in a moist sample which when squeezed will retain its shape when handled freely. Loams are usually soft to the feel.

Sandy Clay Loam: A sandy clay loam is like a clay loam but sand grains can be felt.

Silty Clay Loam: A silty clay loam is like a clay loam but silty as well and smooth to the touch.

Clay Loam: More easily moulded into a shape than a loam, a clay loam rolls out to a thin ribbon between the palms while a loam will break-up. When dry a clay loam will form a lump but is not as tough to break as a clay.

Sandy Clay: A sandy clay is like a clay but sand grains can be felt.

Silty Clay: A silty clay is like a clay but smoother.

Clay: Clays are tough and can be moulded into shapes when moist. Clays form a long flexible ribbon when rubbed between the palms and the ribbon can often be bent into a "U" shape without breaking. Clays dry into very hard clods.

It should always be remembered that soil texture often varies with depth and that the properties of the topsoil are affected by the properties of the subsoil.

Structure

Structure is the arrangement of soil particles and the pore spaces between them. A soil with structure which is beneficial to plant growth has stable aggregates between 0.5 and 2 mm in diameter. Such soils have good aeration and drainage.

Chemical Properties

The inorganic minerals of soils consist primarily of silicon, iron and aluminium which do not contribute greatly to the nutritional needs of plants. Those in the clay fraction have the capacity to retain nutrients in forms which are potentially available for plants to use.

Organic Component

The organic matter of soil usually makes up less than 10% by weight of soil. It can be subdivided into the living and the non-living fractions. The non-living fraction contributes to the soil's ability to retain water and some nutrients and to the formation of stable aggregates.

Organic Matter

The organic matter fraction of soils comes from the decomposition of animal or plant products such as faeces and leaves. Soil organic matter contributes to stable soil aggregates by binding soil particles together.

Plants living in soil continually add organic matter in the form of roots and debris. Decomposition of this organic matter by microbial activity releases nutrients for the growth of other plants.

The organic matter content of a soil depends on the rates of organic matter addition and decomposition. Soil micro-organisms are the primary agents responsible for the decomposition of organic matter such as plant residues. Initially, the sugars, starch and certain proteins are readily attacked by a number of different micro-organisms. The more resistant structural components of the cell wall are decomposed relatively slowly. The less easily decomposed compounds, such as lignin and tannin, impart a dark colour to soils containing a significant organic matter content.

The decomposition rate of organic materials depends on how favourable the soil environment is for microbial activity. Higher decomposition rates occur where there are warm, moist conditions, good aeration, a favourable ratio of nutrients, a pH near neutral and freedom from toxic compounds.

Soil Organisms

The soil contains numerous organisms ranging from microscopic bacteria to large soil animals such as earthworms. The soil micro-organisms include bacteria, fungi, actinomycetes, algae, protozoa and nematodes.

The diversity of soil organisms can both assist and hinder plant growth. Beneficial activities include organic matter decomposition, nitrogen fixation, transformation of essential elements from one form to another, improvement in soil structure through soil aggregation, and improved drainage and aeration.

- Under some circumstances soil organisms compete with plants for nutrients.

- Bacteria are the smallest and most numerous micro-organisms in the soil.

- They make an important contribution to organic matter decomposition, nitrogen fixation and the transformation of nitrogen and sulphur.

The fungi and actinomycetes contribute beneficially to organic matter decomposition. The group of large soil animals includes earthworms, which incorporate organic matter into the soil as well as improving aeration and drainage by means of their channels. Some soil fungi, nematodes, and insects feed on roots and lateral shoots to the detriment of plants.

Physical and Chemical Properties of Soil

Physical Properties of Soils

Physical properties of the soil can be discussed under the following heads:

1. Soil separates and texture,

2. Structure of soil,

3. Weight and soil density,

4. Porosity of soil,

5. Permeability of soil,

6. Soil colour,

7. Temperature of soil, and

8. Soil Plasticity, Compressibility and Erodibility.

Soil Separates and Soil Textures

Mineral fraction of soil consists of particles of various sizes. According to their size, soil particles are grouped into the following types.

Table: Soil separates

Name of soil particle	Diameter range of soil particle in millimetre
1. Coarse particles or gravels	More than 2.00
2. Coarse sands	2.00 – 0.20
3. Fine sands	0.20 – 0.02
4. Silts	0.02 – 0.002
5. Clays	Below 0.0002

The particle sizes of above groups are suggested by International Society of Soil Science. In India, international system of particle differentiation is commonly followed. The particle types are generally called 'soil separates' or 'soil fractions'. Amount of soil separates is determined by a process known as mechanical analysis. In this process, soil sample is crushed and screened through a 2 mm round hole sieve. The screened soil is then homogeneously dispersed in water and allowed to settle.

In suspension, particles of largest dimensions will settle first and those of smaller dimensions will settle afterwards. Individual soil separates are identified on the basis of their respective diameter ranges. Soil separates (sand, silt and clay) differ not only in their sizes but also in their bearing on some of the important factors affecting plant growth, such as, soil aeration, workability, movement and availability of water and nutrients.

Important characteristics of different soil separates are as follows:

Sand

This fraction of soil consists of loose and friable particles of 2.203—.02 mm diameter. Sand particles can be seen by unaided eye. These particles, although inactive, constitute the framework of the soil. They play less important role in physicochemical activities. When coated with clay, these sand particles take very active part in chemical reactions. Sands increase the size of pore spaces between soil particles and thus, facilitate the movement of air and water in the soil.

Silt

It consists of soil particles of intermediate sizes between sand and clay (diam range. 02—.002 mm). Silt, when wet, feels plastic but in dry state feels like flour or talcum. Coarse silt shows little physicochemical activities but finer grades play important role in some chemical processes. Silty soil has got larger exposed surface area than the sandy soil. Silty soils contain sufficient quantities of nutrients, both organic and inorganic. That is why they are very fertile. Soils rich in silt possess high water holding capacity. Such soils are good for agriculture.

Clay

This soil fraction contains smaller particles than silt (below. 002 mm diameter) which exhibit plasticity and smoothness when wet and hardness when dry. Owing to their smallest size and colloidal nature, the clay particles expose extremely large surface area. They take very active part in physicochemical reactions of the soil. Clay soils have fine pores, poor drainage and aeration and thus they have highest water holding capacity. The clay acts as store house for water and nutrients.

Some soils are fine, while others are coarse. It is so because of the fact that the relative percentage of sand, silt and clay differ from soil to soil. The relative percentage of soil separates of a given soil is referred to as soil texture. Texture of soil for a given horizon is almost a permanent character, because it remains unchanged over a long period of time.

The relative percentages of soil separates of average samples are almost infinite in possible combinations. It is, therefore, necessary to establish limits of variations among soil fractions so as to group them into textural classes. The common textural classes, as recognized by USDA (U.S. Department of Agriculture) are given in the following table. These classes are recognized on the basis of relative percentage of separates; sand, silt and clay.

Table: Soil textural classes

Soil classes or	Range in relative percentages of soil separates		
Textural name	Sand	Silt	Clay
Sandy soil	85 – 100	0 – 15	0 – 20
Loamy sand	70 – 90	0 – 30	0 – 15
Sandy Loam	43 – 80	0 – 50	0 – 20
Loam	23 – 52	28 – 50	7 – 27
Silt loam	0 – 50	50 – 88	0 – 27
Silt	0 – 20	88 – 100	0 – 12
Sandy clay loam	45 – 80	0 – 28	20 – 35
Clay loam	20 – 45	15 – 53	27 – 40
Silty clay loam	0 – 20	40 – 73	27 – 40
Sandy clay	45 – 65	0 – 20	35 – 45
Silty clay	0 – 20	40 – 60	40 – 60
Clay	0 – 45	0 – 40	40 – 100

This chart is adapted from fraction system of U.S.D.A. If relative percentages of soil separates are known, the soil can be given textural name. For this purpose equilateral triangles are used. The most widely used Equilateral triangles are international equilateral triangle and the one used by USDA. These consist of three angles and its area is divided into twelve groups representing twelve different textural classes. Each group covers definite range of percentages of sand, silt, and clay. In the triangles, left side line represents the clay %, right side line represents percentage of silt and base represents percentage of sand.

Each arm of the triangle is divided into ten divisions representing soil separate's percentage. These divisions are further divided into ten small divisions; each small division represents one per cent of soil separate. The percentages of sand, silt, and clay obtained after mechanical analysis of the given soil are read on the equilateral triangle.

In using the diagram as indicated the percentages of silt and clay should be located on silt and clay lines respectively. The line in case of silt is then projected inward parallel to clay side of the triangle and in case of clay it should be projected parallel to the sand side. The three lines; one representing sand percentage, other representing silt percentage and the third clay percentage meet at a point in the triangle. The compartment in which the point falls indicates textural name for the given soil sample. The knowledge of soil texture is of great help in the classification of soil and in determination of degree of weathering of rock.

Structure of Soil

Sand, silt and clay are found in aggregated form. Arrangement of these soil particles on certain defined patterns is called soil structure. The natural aggregates of soil particles are clod peds whereas an artificially formed soil mass is called clod. Ped differs from fragment because the latter refers to the broken ped. Ped differs from concretion in the sense that the latter is formedinthe soil by precipitation of salts dissolved in percolating water.

Soil structure also reveals the colour, texture and chemical composition of soil aggregates. Soil structure is influenced by air moisture, organic matter, micro-organisms and root growth. When many particles or peds are aggregated into cluster, a compound particle is formed.

Soil Structure is described under the following three categories:

Type

This indicates the shapes or forms and arrangement of peds. Peds may be of various shapes, such as granular, crumb, angular blocky, sub angular blocky, platy and prismatic. Different types of peds and their properties are described in table.

Type of ped	Properties
(1) Granular	Small, spheroidal and nonporous
(2) Crumb	Small, porous and spheroidal
(3) Angular blocky	Block-like with sharp ends, one end may be pointed.
(4) Subangular blocky	Block-like but bounded by other aggregates.
(5) Platy	Plate like, sometimes plates are overlapped.
(6) Prismatic or Columnar	Prism like but without rounded surface.

Size Class

These are as follows:

1. Very fine or very thin

2. Fine or thin

3. Medium

4. Coarse or thick

5. Very coarse or very thick

Grade

It is described under the following four categories:

(i) Structure less

Peds not distinct, i.e., cement or sand like condition.

(ii) Weak

Peds distinct and rarely durable.

(iii) Moderate

Peds moderately well developed, fairly durable and distinct.

(iv) Strong

Peds well developed, quite durable and distinct.

Density and Soil Weight

Density of soil is the mass per unit volume. It is expressed in terms of gm per cubic centimeter. Average density of the soil is 2.65 gms per cubic centimeter. Density of soil varies greatly depending upon the degree of weathering.

For this reason soil density is expressed in two generally accepted forms:

- Particle density or true density; and
- Bulk density.

(i) Particle density: Density of solid portion of soil is called particle density. It is sum total of densities of individual organic and inorganic particles. Average particle density of organic soil varies from 1.2 to 1.7 gms per ml. and that of inorganic fraction varies from 2.6 to 2 78 gms/ ml. Particle density may be calculated as: weight of solids/volume of soils. Particle density divided by density of water gives the specific gravity or relative weight number. Specific gravity of soil particles = Particle density/density of water.

(ii) Bulk density or apparent density: Dry weight of unit volume of soil inclusive of pore spaces IS called bulk density. It is expressed in terms of gm per ml or lbs per cubic foot. It is lesser than the particle density of the soil. Bulk density of soil may be calculated as: weight of soil/volume of soil.

Bulk density of the soil divided by density of water gives volume weight or apparent specific gravity of soil. Bulk density of soil changes with the change in total pore space present in the soil and it gives a good estimate of the porosity of soil. Average density of soil in bulk is 1 5 gm/ml. Organic soils have low bulk density as compared to mineral soils. Soil weight varies in relation to textural classes. Average weight of loam or sandy soil is 80—110 pounds/cubic foot but that of clay ranges between 70 and 100 pounds/cubic foot.

Porosity of Soil

The spaces occupied by air and water between particles in a given volume of soil are called pore spaces. The percentage of soil volume occupied by pore space or by the interstitial spaces is called porosity of the soil. It depends upon the texture, structure, compactness and organic content of the soil. Porosity of the soil increases with the increase in the percentage of organic matter in the soil. Porosity of soil also decreases as the soil particles become much smaller in their dimension because of decrease in pore spaces.

It also decreases with depth of the soil. The pore spaces are responsible for better plant growth because they contain enough air and moisture. Percentage of solids in soils can be determined by comparing bulk density and particle density and multiplying by hundred.

$$\text{Percentage of solids} = \frac{\text{bulk density}}{\text{particle density}} \times 100$$

This Percentage of solids of solids subtracted from total volume (100%) will give the percentage of pore space. Hence, the formula :

$$100\% - \left[\frac{\text{bulk density}}{\text{particle density}} \times 100 \right] = \text{Percentage of pore space}$$

Depending upon the size pore spaces fall into two categories.

These are:

1. Micro-pore spaces (capillary pore spaces).

2. Macro-pore spapes (non-capillary pore spaces).

Capillary pore spaces can hold more water and restrict the free movement of water and air in soil to a considerable extent, whereas macro-pore spaces have little water holding capacity and allow free movement of moisture and air in the soil under normal conditions.

Permeability of Soil

The characteristic of soil that determines the movement of water through pore spaces is referred to as soil permeability. Soil permeability, because it is directly dependent on the pore size, will be higher for the soil with large number of macro-pore spaces than that for compact soil with a large number of micro-pore spaces (capillary spaces). Permeability of soil also varies with moisture status and usually decreases with the gradual desiccation of soil. In the arid regions, groundwater moves upwardly through capillary action and bring sodium, potassium and calcium salts with it in dissolved state on the surface of soil. The water evaporates and inorganic salts precipitate on the surface of the soil. As a result of this, the soil becomes less permeable and the productive capacity of soil is reduced.

Soil Colour

Soils exhibit a variety of colours. Soil colour may be inherited from the parental material (Le., lithochromic) or sometimes it may be due to soil forming processes (acquired or genetic colour). The variations in the soil colour are due to organic substances, iron compounds, silica, lime and other inorganic compounds.

The organic substances impart black or dark greyish-black colour to the soil. Iron compounds are responsible for brown, red and yellow colours of soils. Iron oxides in combination with organic substances impart brown colour which is most common soil colour. Silica, lime and some other inorganic compounds give light white and grey tinges to the soil.

Soil colour influences greatly the soil temperature. The dark coloured soils absorb heat mort readily than light coloured soils. black cotton soil absorbed 86% of the total solar radiations falling on the soil surface as against 40% by the grey alluvial soil. Soil colour is used as an important criterion for description and classification of soil. Many soils are named after their prominent colours, such as black cotton soil, red-yellow latosol, grey hydromorphic soils and so on.

Soil Temperature

The chief sources of soil heat are solar radiations and heat generated in the decomposition of dead organic matters in the soil and heat formed in the interior of earth. The soil temperature greatly affects the physico-chemical and biological processes of the soil. Temperature of soil depends upon the temperature of atmospheric air and on moisture content. It is controlled by climate, colour of soil, slope, and altitude of the land and also by vegetational cover of the soil.

The average annual temperature of soil is generally higher than that of its surrounding atmosphere. Ramdas and M.S. Katti recorded surface temperature of black cotton soil as high as 165″ at Poona (Maharashtra), India. Surface temperature of soil shows considerable fluctuations but soil temperature below certain depth remains more or less constant and is not affected by diurnal or regional temperature changes.

Studies made by Leather at Pusa Research Institute in Bihar (India) showed that diurnal temperature difference at the level 12 inches below the soil surface was only 1 °C and at a depth of 24″ it seldom exceeded 0.1 °C. At the depth of 3 or 4 feet, the temperature remains almost constant.

Soil Plasticity, Compressibility and Erodibility

Soil plasticity is a property that enables the moist soil to change shape when some force is applied over it and to retain this shape even after the removal of the force from it. The plasticity of soil depends on the cohesion and adhesion of soil materials. Cohesion refers to the attraction of substances of like characteristics, such as, that of one water molecule for another. Adhesion refers to the attraction of substances of unlike characteristics. Soil consistency depends on the texture and amount of inorganic and organic colloids, structure and moisture contents of soil.

Compressibility

It refers to the tendency of soil to consolidate or decrease in volume. The compressibility is partly a function of elastic nature of soil particles and is directly related to settlement of structures. With the decrease in the moisture contents soils gradually tend to become less sticky and less plastic and finally they become hard and coherent. Plastic soils have great cohesion force. It is only because of cohesion property the moist clay soils frequently develop cracks when they become dried. Coarse materials such as gravels and sands have low compressibility and the settlement is considerably less in these materials as compared to highly compressible fine grained organic soils.

Erodibility

It refers to the ease with which soil materials can be removed by wind or water. Easily eroded materials include unprotected silt, sand and other loosely consolidated materials, Cohesive soils (with more than 20% clay) and naturally cemented soils are not easily removed from its place by wind or water and, therefore, have a low erosion factor.

Chemical Properties of Soils

Chemical properties of soils can be described under the following heads:

1. Inorganic matters of soil,

2. Organic matters in soil,

3. Colloidal properties of soil particles,

4. Soil reactions and Buffering action:

    a. Acidic soils

    b. Basic soils

Inorganic Matters of Soil

From the accounts given in the description of weathering process it is clear that compounds of aluminium, silicon, calcium, magnesium, iron, potassium and sodium are chief inorganic constituents of soils. Besides these, the soils also contain small quantities of several other inorganic compounds, such as those of boron, magnesium, copper, zinc, molybdenum, cobalt, iodine, fluorine etc. The amounts of these chemicals vary in soils of different places. Chemical composition of soil of one horizon differs greatly from the composition of soil in the other horizon.

Organic Matters in Soil

Organic component of the soil consists of substances of organic origin; living and dead. In sandy soil of arid zone, it is found in very poor quantity (one or less than one per cent) but in peaty soil, it may be as high as 90%. When the plants and animals die, their dead remains are subjected to decomposition.

As a result of decomposition a number of different organic products or compounds are formed from the original residues. In the course of decomposition, the original materials are converted into dark coloured organic complexes, called humus. Sometimes living micro-organisms add sufficient amount of organic matters in soil in the form of metabolic wastes.

Chemists have been attempting to unravel the details of humus composition since the earliest days of soil science, and have got much success but more is yet to be discovered. In terms of specific elements, the organic component of soil contains compounds of carbon, hydrogen, oxygen, phosphorus, nitrogen, sulphur and small amount of other elements also. Only small fraction of total organic matter is soluble in water but majority of them are soluble in alkali solution.

Chemically humus contains the following organic molecules:

A. Amino Acids:

1. Glutamic acid

2. Alanine

3. Valine

4. Proline

5. Cystine

6. Phenyl alanine

B. Proteins:

1. Purines

   a. Guanine

   b. Adenine

2. Pyramidines

    a. Cytosine

    b. Thymine

    c. Uracil

C. Aromatic Molecules:

D. Uronic Acids:

1. Glucuronic acid

2. Galacturonic acid

3. Lactic acid

E. Aliphatic Acids:

1. Acetic acid

2. Formic acid

3. Succinic acid

F. Aminosugars:

1. Glucosamine

2. N. Acetylglucosamine

G. Pentose Sugars:

1. Xylose

2. Arabinose

3. Ribose

H. Hexose Sugars:

1. Glucose

2. Galactose

3. Manose

I. Sugar Alcohols:

1. Inosital

2. Mannitol

J. Methyl Sugars:

1. Rhamnose

2. Fucose

3. 2-0, methyl D-xylose

4. 2-0, methyl D-arabinose

Besides these compounds locked up in the humus fraction, the soil also contains fats, oils, waxes, resins, tannin, lignin and some pigments.

Waksman and Stevans have proposed the following method for separation of different organic compounds present in the soil:

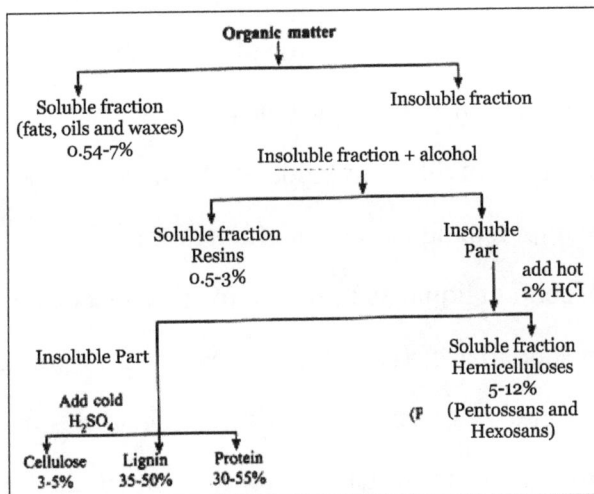

Another modified method for separation of the various organic compounds from the soil is as follows:

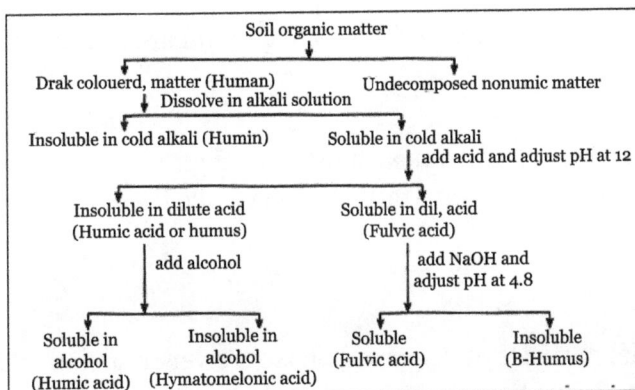

The fractions are not pure chemical compounds but are in the form of mixtures of several substances. They are found in colloidal state in the soil.

Colloidal Properties of Soil Particles

There are two types of substances namely crystalloids and colloids:

- Crystalloids are those crystalline solid substances which form true solution on being mixed with other substances. In true solution, crystal particles cannot be seen with the help of microscope.

- Colloid is really speaking amorphous state of the substances which do not form true solution if mixed with other substances. The particles of colloidal substances float in the solvent in suspension state but do not tend to settle at the bottom. Colloids are not found in ionic or molecular form but are found in aggregates of atoms or molecules.

Colloidal system or suspension contains two phases which are:

1. Dispersion phase, i.e., medium in which the particles are suspended,

2. Dispersed phase, i.e., suspended particles.

Colloidal suspension may be of different kinds, such as:

1. Suspension of liquid, in liquid, as milk (fats in water).

2. Suspension of solid in liquid as India ink (or clay suspension in water).

3. Suspension of solid in gas, as smoke (coal particles suspended in air).

4. Suspension of liquid in gas, e.g., cloud and fogs in atmosphere.

The commonest colloids are those which remain suspended in a liquid medium. If the colloidal suspension exhibits properties of fluid, it is called sol, but sometimes sols exhibit solid like behaviour and form solid or nearly so. This condition is called gel. Some sols form reversible gel while the others form irreversible gel.

Some of the important properties of colloids in general are as follows:

1. Particle size

Crystalloids and colloids differ from each other in their size range Particles of crystalloid in true solution are 0.2 to 1 mμ (millimicron) while those of colloids in suspension are 1 to 200 mμ.

2. Adsorption

Because the colloidal particles of dispersed phase are very small, they have got large exposed surface areas. Owing to their large exposed surface areas, these colloidal particles show great adsorptive capacity. In adsorption, particles of particular substances come to lie on the surface of colloids and they do not enter deep in the colloidal particles.

3. Electrical Properties

The electrically charged colloids are termed as micelles. Colloids have some electrical charge on them. They may be charged either positively or negatively. Colloidal particles of one electrical charge have tendency to attract colloids of opposite charge. In the soil clay particles are negatively charged, thus they attract cations (+ charged ions).

Colloid particles differ from electrolytes in the fact that when electric current is passed in the colloidal suspension, all the colloidal particles are attracted towards one electrode or the other depending upon the nature of charge they carry on them. This phenomenon is called electrophoresis.

The electrolytes, when dissolved in solvent dissociate into two types of ions among which half will bear positive charge (cations) and remaining half will bear negative charge (anions) When electric current is passed in the solution of electrolyte all the positively charged ions will accumulate on negative pole and remaining negatively charged ions will collect on positive pole.

4. Coagulation or flocculation of colloidal particles

Colloidal particles in the suspension can be coagulated either by heating or by adding some substances which contain opposite charged ions. When substances carrying positive ions are added in suspension containing negatively charged colloid particles, ions will move and accumulate on the surface of colloids carrying opposite charge. Finally a stage comes when colloidal particles cannot attract more opposite charged ions, this is called isoelectric point. As a result of ion accumulation on their surface, the colloids first become large and heavier and finally they tend to settle at the bottom in floccules. This process is known as flocculation.

5. Tyndal phenomenon

Colloidal particles in suspension can be seen when a strong beam of light is passed through suspension and observer looks it from the place at right angle to the path of light. The colloidal particles become visible as strongly illuminated particles and they appear bigger than normal size. This phenomenon is known as "Tyndal effect".

6. Brownian movement

Colloidal particles when suspended in dispersion medium show a characteristic continuous zig-zag motion, called Brownian movement. This type of movement was first observed by English botanist Robert Brown, hence it is called Brownian movement. The movement is exhibited because of characteristic collision of one particle with others. This prevents the particles from settling down.

7. Dialysis

Because colloidal particles in suspension are larger than the particles of crystalloid in true solution and are larger than the diameter of pores of porous membranes, e.g., parchment membrane, they are not allowed to be filtered down and are retained the membrane. Thus, they can be separated in pure state from the mixture of crystalloids and colloids by filtration process. This separation process is known as dialysis.

Colloidal Fraction of Soil

There are two types of colloids in the soil. These are:

1. Mineral colloids or clay colloids, and

2. Organic or humus colloids.

These two colloidal fractions of soil are very intimate to each other and it is very difficult to separate them. The inorganic colloids occur as very fine particles and organic colloids occur in the form of humus particles. The soil colloid particles show almost all the characteristics of typical colloidal system, i.e., adsorption, Tyndal effect, Brownian movement, coagulation, electrophoresis, dialysis etc.

Clay Colloids

As regards the size, clay fraction of soil contains both non-colloidal and colloidal particles. Some clay particles may be as large as .002 mm in diameter but some may be smaller than normal colloid size (normal size of colloid particle is from 1 to 200 mμ). The clay particles are formed mainly of silica, alumina, iron and combined water. Colloidal clay may also contain rich accumulation of plant nutrients.

Early researchers of soil science have described clay colloids as spherical particles and their sizes were mentioned in terms of their diameters, but recent electron micrographs reveal that particles occur in layers or plates and each c ay particle appears as if it is composed of a large number of plate-like units. These units or flakes of clay are held together by a force of attraction. The plate-like clay particles expose large surface area on which moisture and cations (+ ions) are held. The finer the clay particles the greater will be the percentage of hygroscopic moisture.

If clay is suspended in distilled water, shaken, and then a little NH_4OH is added to suspension and allowed to settle, after a few minutes large particles settle down but finer particles remain in suspended state. When a little limewater is added to suspension, fine suspended particles increase in size and form small floccules which have a tendency to settle down. Thus, finer clay particles show flocculation property. Now if some acid is added, the floccules are broken and the clay particles will return to their normal size. This process is known as deflocculating.

The clay particles are negatively charged, hence they can hold thousands of positively charged ions of mineral nutrients on their surfaces. The clay colloids are lyophilic (water loving). So, they are important from the standpoint of the adsorption of large quantity of water (perhaps, 5-10 layers of water molecules are held on the surface of clay colloid).

Organic Colloids or Humus Colloids

Organic colloids in the soil are chiefly due to presence of humus. The humus contains 8% each of lignin, protein, polyuronides (sugars and uronic acid complex). Organic colloids may be present in appreciable proportion in the soils. In sandy soil, it forms minor part of colloids. In peaty soil, organic colloids may be more than 50%. These colloids show adsorptive capacity many times greater than clay colloids. Organic colloids are negatively charged like clay colloids. Addition of organic colloids to the sandy soil increases temporarily its moisture and nutrient retaining capacity.

Cation Exchange

Since the soil colloids (clay and organic colloids) have negative charges on them, they attract and hold positive ions (cations). When cations are added to the soils such as Ca^{++} in the form of lime, K^+ ions in the form of potassium fertilizer, and NH^{++} in the form of ammonium fertilizer, the adsorption of cations will take place on the surface of colloid micelle and this will be accompanied by release of one or more ions held by colloid micelle.

This is known as cationic exchange. For example, suppose that colloid micelle has one half of its capacity satisfied with Ca^{++} ions, one quarter with K^+ ion and remaining one quarter with H^+ ions. Now the colloid is treated with KCl solution. The K^+ ions will first replace Ca^{++} ions and then H+ ions. The Ca^{++} and H^+ ions will combine with Cl^- ions of KCl and will form $CaCl_2$ and HCl respectively.

The cation exchange in the soil may take place between:

1. Cations present in the soil solutions and those already present on surface of soil colloids,

2. Cations released by plant roots and those present on the surface of soil colloids,

3. Cations present on the surface of two clay crystals either two organic colloids or an organic colloid and a clay colloid.

Exchange reaction is very quick and reversible and the exchange of ions continues till equilibrium is attained. All cations are not adsorbed with equal ease. Some are easily adsorbed while others are replaced with difficulty. Divalent cations are more effective than the monovalent ones. Hydrogen is exception because it is held by colloids most tenaciously and it is most powerful replacer of cations.

Replacing capacities of some cations are compared here,

$$H^+ > Ca^{++} > K^+ > Na^+ > Na^+$$

The number of cations adsorbed per unit weight of one hundred grams dry soil is called cation exchange capacity. More scientifically, cation exchange capacity of soil is the sum total of exchangeable cations adsorbed per unit weight of one hundred gms of dry soil.

Factors which are responsible for cation exchange or base exchange are as follows:

1. Relative concentration and number of cations present in the soil,

2. Replacing capacity of the ions,

3. Number of charges on the ions.

Anion Exchange

Soils rich in organic colloids show anion exchange also. In this process, negatively charged ions held by colloids are replaced by OH^-, $H_2PO_4^-$, SO_4^-, and $NO3^-$ ions. The relative order of exchange is,

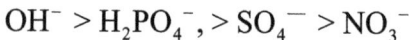

$$OH^- > H_2PO_4^-, > SO_4^- > NO_3^-$$

Among these anions, exchange of PO_4^-, ions is most important. SO_4^- and NO^- are not retained in the soil for long period of time, hence not available for anionic exchange. Laterite soils have high adsorptive and fixation capacity for PO_4^- than black soils.

Anionic and cationic exchange reactions are important in agriculture. Several soil scientists have shown that the capacity of soil to exchange cations is the best index of soil fertility. The predominance of desirable ions in the exchange complex brings about good physical cations and favorably influences the microbial activities in the soil, such as ammonification nitrification, etc. The knowledge of cation and anion exchange is of great help in reclaiming acidic and saline or alkaline soils.

Anionic and cationic exchange reactions are important in agriculture. Several soil scientists have shown that the capacity of soil to exchange cations is the best index of soil fertility. The predominance of desirable ions in the exchange complex brings about good physical cations and favorably influences the microbial activities in the soil, such as ammonification nitrification, etc. The knowledge of cation and anion exchange is of great help in reclaiming acidic and saline or alkaline soils.

Soil Reaction

Many chemical properties of soils centre round the soil reaction. As regards their nature, some soils are neutral, some are acidic and some basic. The acidity, alkalinity and neutrality of soils are described in terms of hydrogen ion concentrations or pH values. In order to understand soil reaction, the knowledge of pH is very necessary.

It can be understood in the following ways:

Water dissociates into H^+ ion and OH^- ion. Hence the ionic constant of water can be represented as follows:

Ionic constant of water $= [H^+][OH^-]/[H_2O]$. But the rate of dissociation of water is so slow that ionization constant of water can be expressed simply as product of concentration of H^+ and OH^- ions, thus ionization constant of water $Kw = [H^+][OH^-]$. Concentration of H^+ and OH^- ions are expressed in terms of equivalents per litre. Only one molecule in ten million water molecules is in dissociated condition.

At neutrality, H^+ concentration is 0.0000001 or 10^{-7} gm of hydrogen per litre solution. The ionization constant of water is 10^{-14} at 25 °C and thus in any aqueous system products of H^+ and OH^- ion concentration is 10^{-14}.

Now, the above equation can be written as:

$$10^{-14} = [H^+][OH^-]$$

This can also be represented in the following way by dividing both sides in one and taking logarithms.

Hence,

$$\log \frac{1}{[Kw]} = \log \frac{1}{[OH^-]} + \log \frac{1}{[H^+]} = 14$$

The value of log 1/[H$^+$] and log 1/[OH$^-$] are generally pH and poH respectively. These pH and pOH are indices of the acidity and alkalinity respectively. Thus, pH can be defined as negative logarithms of the H^+ ion concentration. When the system is neutral, pH will be equal to pOH and when Kw is 10^{-14}, the value of pH and pOH at neutral point will be 7 for each. When pH value is less than 7, it is acidic. The pH value above 7 indicates alkalinity. Ifina system, hydrogen ion concentration is 1/.000001 or 000001 gm/litre, the pH value will be 6.

It is more clear from the following calculation:

$$pH = \log \frac{1}{[H^+]} = \log \frac{1}{0.000001} \text{ or } \log 10,00,000 = 6.$$

Thus at pH value of 6, the H^+ ion concentration is increased 10 times than the H^+ concentration at pH value of 7. Like this, at every lower pint H^+ ion concentration will increase by a multiple of 10. As the product of H^+ ion and OH^- ion concentrations at neutrality is always 10^{-14}, the pOH can be determined from pH value. Suppose, pH value of a solution is 6, the pOH value will be 8 ($10^{-6} + 10^{-8} = 10^{-14}$) pH scale is divided into

14 divisions or pH units from 1 to 14. Soil with pH value of 7 is neutral, that below pH 7 is acidic and that with pH value above 7 is alkaline.

From the pH value intensity of acidity in the soil is expressed but these values are not the measure of total acidity because they do not indicate the reserved acidity or relative acidity. For example, there are two soil samples which have similar pH values but they require different quantities of lime for neutralization. It means that the quantities of acids are different in the given weight of above two soils.

Buffer Action

It refers to the resistance to change in pH of a system. Such solutions as are reasonably permanent in pH value even after addition of some alkali or acid to them are called solution with reserved acidity or alkalinity or more often "buffer solutions". Suppose, a certain amount of acid is added to distilled water, the resulting solution will show acidic reaction and that will have a pH below 7, but if the same quantity of acid is added to a neutral soil suspension there would be very minor change in pH. This property of soil to resist a change in pH is called "buffer action".

Buffer solutions are usually formed of a mixture of salt of weak acid and acid itself in various proportions, as for example, a mixture of sodium acetate and acetic acid if added to water will result in a buffer solution. In the mixture solution we have mainly sodium ions and acetate ions. In water there will be some H^+ and OH^- ions. In buffer solution, acetate ions are in excess, owing to presence of well ionised sodium acetate. If H^+ ions are added to this solutions they will combine with acetate ions to give acetic acid of low ionisation power.

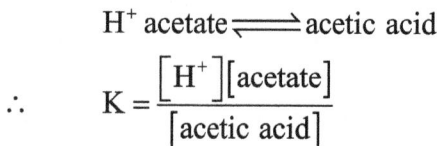

$$H^+ \text{ acetate} \rightleftharpoons \text{acetic acid}$$

$$\therefore \quad K = \frac{\left[H^+\right]\left[\text{acetate}\right]}{\left[\text{acetic acid}\right]}$$

Hence, there will be a little increase in pH. The addition of acid to buffer solution then makes little difference in the pH value.

Buffering in Soil

Soils should have good buffering capacity. Therefore, it is necessary to add considerably large amount of acids or alkalis in order to bring about any change in the original pH of soil. Buffering action is due to presence of large quantity of weak acids and their salts in the soil. Phosphates, carbonates, bicarbonates and other salts of weak inorganic acids and corresponding acids themselves are important buffering agents in the soils. Besides these, colloids associated with cations are important buffering agents. The buffering action of soil is directly governed by the amount and nature of clay and organic or humus colloids present in it.

Buffering action of soil is important in agriculture in the following respects:

(1) Stabilization of pH: This protects the higher plants and micro-organisms form direct adverse and injurious effects of sudden change in soil reaction.

(2) Amount of amendments necessary to correct the soil reaction: The greater is the buffering capacity of soil the smaller will be the amount of the amendments required such as lime, sulphur etc. to correct the acidity or alkalinity.

Soil Fertility

Soil fertility is the sustainable capacity of a soil to produce good yields of high quality on the basis of chemical, physical and biological factors. The three factors are needed to assess soil quality or fertility. Much work has been done on physical, chemical, climatic and cultivation data. But biological soil components have never been used to assess soil quality.

The parameters for estimation of biological factors are soil respiration, nitrogen mineralisation, nitrification, denitrification, nitrogen fixation, soil enzyme activity (e.g. urease, acid phosphatase and alkaline phosphatase, etc.).

The level of microbial biomass of a soil denotes soil quality at a time on the basis of above parameters. The figure below shows the integrated evaluation of biological, chemical, physical, cultivation and climatic data.

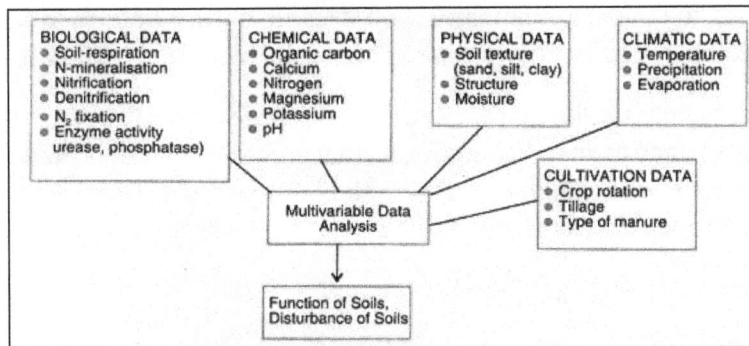

Integrated evaluation of biological, chemical, physical, cultivation and climatic data for a soil which shows functioning of soil. These also act as indicators of disturbances arising in soil.

Functions of Soil

The basic natural function of soil system should be:

1. Primary production by creation a rooting substrate,

2. To support above ground,

3. To decompose organic material resulting in release of simple organic compounds and minerals,

4. To exchange gases, solutes and organisms between the aqueous, solid and gas phases,

5. To support diversity of live forms and their adaptations, in response to heterogeneous physical, chemical and biological conditions.

Microbial Biomass as an Index of Soil Fertility

In recent years, the role of microorganisms in decomposition of organic matter and mineralization of bio-elements, and more specifically the large pool of mineral nutrients in their cytoplasm and their turn over into the soil again available to plant roots, have been much studied in many countries including India.

Microbial biomass, irrespective of soil types, is a part of soil organic matter which includes the living microbial components i.e. bacteria, actinomycetes, fungi, algae, protozoa and other microfauna. Microbial biomass is influenced by affecting the microbial activity. The microbial biomass in nutrient poor system acts as a sink and a source of nutrients.

It accumulates during dry period and releases these rapidly during monsoon period so that plant growth can be started. This accumulation of nutrients is called immobilization. Thus in nutrient-poor ecosystem, nutrient retention and withdrawal mechanisms are effective.

Cultivation leads to considerable loss of soil organic matter and, therefore, microbial biomass also changes. The turn over time of microbial biomass is estimated at 1-3 years.

Since microbial biomass is the most active fraction of soil organic matter, therefore, any change in organic matter may be supposed that microbial biomass too has changed. microbial biomass is associated with water-stable macro and micro aggregates in forest, savana and cropland soils of a seasonally dry tropical region.

A fertile soil plays an important role in enhanching the yields of tomatoes.

Soil Metagenomics

Metagenomics refers to the use of DNA sequencing to determine the phylogenetic and functional gene complement of a sample, such as microbial community DNA in soil. A shotgun metagenomic approach relies on sequencing of total DNA extracted from a given sample, without prior cloning into a vector.

Soil as a Microbial Habitat

Soil has several unique properties compared to other microbial habitats that are important to considerwhen discussing the topic of soil metagenomics. Although metagenomics is revealing new information about phylogenetic and functional genes in some soils, it is not possible to adopt the information available to date to all soils.

There are different classes of soils that vary according to texture and other geochemical characteristics. The traditional soil classification scheme divides soils into 12 major orders, each of which can be subdivided into suborders and classes. The 12 soil orders are the following:

1. Gelisols (soils with permafrost within 2 m of the surface).

2. Histosols (organic soils).

3. Spodosols (acid forest soils with a subsurface accumulation of metal humus complexes).

4. Andisols (soils formed in volcanic ash).

5. Oxisols (intensely weathered soils of tropical and subtropical environments).

6. Vertisols (clayey soils with high shrink/swell capacity).

7. Aridisols ($CaCO_3$-containing soils of arid environments with subsurface horizon development).

8. Ultisols (strongly leached soils with a subsurface zone of clay accumulation and <35 % base saturation).

9. Mollisols (grassland soils with high base status).

10. Alfisols (moderately leached soils with a subsurface zone of clay accumulation and >35 % base saturation).

11. Inceptisols (soils with weakly developed subsurface horizons).

12. Entisols (soils with little or no morphological development).

The soil properties that are recognized by soil taxonomists are also important for governing microbial community composition and activity. For example, soil pH is known to be a key driver for microbial communities. Therefore, when considering "soil" as an ecosystem, it is important to take into account the type of soil that is being studied.

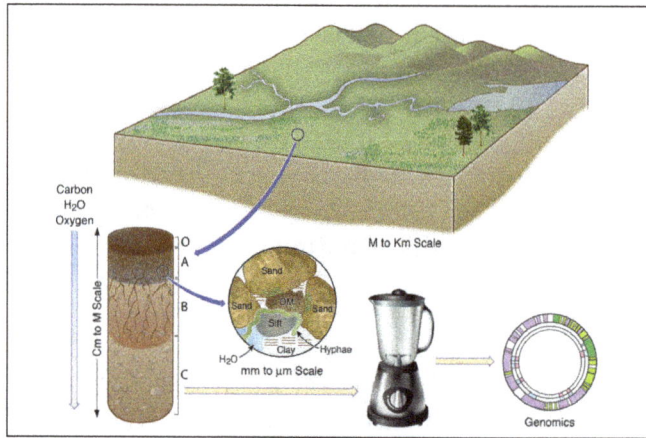

Microheterogeneity of soil

Another unique feature of the soil environment is its spatial heterogeneity at different scales. For example, soil properties can vary considerably across a landscape and are dependent on several features including the plant type, slope of the land, soil moisture, and other aspects of the terrain. The soil profile also changes with depth. Soil is normally classified according to depth horizons with a surface organic "O" horizon that may or may not be present, followed by the "A" horizon and the subsurface "B" horizon that is influenced by plant roots and finally the underlying "C" horizon. Usually, the microbial density and activity are highest at the surface and decrease with depth. Also, the types of microorganisms vary with depth. For example, there can be decreases in oxygen concentrations with depth, resulting in a corresponding increase in anaerobes with depth. Some recent soil metagenomic studies have shown a vertical variation in community structure. The latter two studies compared active layer and deeper permafrost layers that have very different properties.

Different soils also have different mineral compositions and redox conditions and thus different potential electron acceptors. For example, in anaerobic zones, the microbial activity will be dependent on the availability of electron acceptors with decreasing redox potential as follows: $O_2 > NO_3 > Mn > Fe > SO_4 > CO_2$. Metagenome data can provide a clue to the type of prevailing redox conditions in the soil based on the prevalence of genes for reduction of given types of electron acceptors, i.e., methanogenesis, denitrification, sulfate reduction, etc. For example, Mackelprang et al. found a high abundance of functional genes for methanogenesis and denitrification when screening permafrost metagenomes.

The soil habitat is further complicated because of the partitioning of resources into different microscopic niches, for example, soil pores containing water or organic matter. Microbial life in soil is thus often concentrated into discrete locations in soil aggregates. Due to the spatial heterogeneity at a microbial scale, there can be different microbial populations residing close to each other but physically separated by soil grains or air-filled pores. This becomes a complication when using a metagenomic approach to understand the microbial community composition and function. Ideally, one would sequence individual microscopic soil aggregates to determine which populations are present in individual microscopic habitats. However, this is currently beyond the current level of sequencing resolution, although advances in single-cell sequencing technologies may be applicable in this regard. Usually, soil metagenomes are obtained from at least 1 g of soil that has been homogenized and thus the individual microscopic habitats are not resolved and the composite community is analyzed in the sequence data.

Soil Microbial Activity

Due to the microheterogeneity of soil, microorganisms may be more or less active depending on their access to nutrients and other conditions necessary for activity, including the factors mentioned above. The majority of soil microbes are normally in a dormant or quiescent state as they optimize conditions to become available. For example, actinobacteria are known to be persistent survivors in soil that are resistant to desiccation and can withstand long-term starvation conditions. Therefore, actinobacteria are often abundant in 16S surveys of soil samples. However, whether or not they are active is another matter. A metaproteomic survey of a California grassland soil found a high prevalence of proteins corresponding to Bacillus spore proteins in soil, thus emphasizing the importance of this survival strategy.

Roots can be considered nutrient "hot spots" for microbes living in normally nutrient-poor soil conditions. The portion of soil that is directly influenced by roots is known as the rhizosphere. Rhizosphere microbial communities have been found to be different than communities residing in bulk soil. The rhizosphere effect is therefore something that should be considered when sequencing metagenomes from soil with a cover of vegetation. Unless the roots can be completely separated from samples prior to DNA extraction, the sample probably contains microbes that are influenced by the rhizosphere. The same soil sample can contain regions that are not influenced by roots. Therefore, the resulting metagenome will be a composite of different microbial communities that may be more or less active. When mining the metagenome sequence data for functional genes, the relative amounts of genes involved in activities expected in the rhizosphere, including quorum sensing, nitrogen fixation, etc., will be dependent on the relative amounts of DNA from rhizosphere soil in the sample.

The variable status of microbial activity in soil is a complication when analyzing soil metagenome sequence data that can include DNA extracted from microbes in different physiological states, ranging from active and growing to dormant or even dead. One option could be to fractionate the soil microbial community according to their physiological status prior to metagenome sequencing. For example, specific members of the soil community could be enriched with nutrients to increase the fraction of the community that is specifically capable of growth and metabolism of the added nutrients prior to DNA extraction. If a specific ^{13}C-labeled substrate is added, the DNA from microbes that incorporate the ^{13}C label during metabolism of the substrate can be fractionated on density gradients, a technique commonly referred to as stable isotope probing (SIP). This approach was used to enrich methanotrophs in a forest soil by incubation with ^{13}C-labeled methane. The ^{13}C-labeled DNA was cloned into a BAC library and screened for genes involved in methane oxidation.

Another option is to add bromodeoxyuridine (BrdU) as a thymidine analogue that is incorporated into the DNA of replicating cells. The DNA with BrdU incorporated can be selectively extracted using magnetic beads coated with antibodies targeted to BrdU. This DNA should then correspond to the growing members of the community. Although not all soil microbes take up BrdU with equal efficiencies, those that do so can be identified as growing using this approach. Sorting of cells prior to DNA extraction has also been proposed as a way to select for cells in a specific physiological state. For example, fluorescence-activated cell sorting (FACS) can be used to distinguish cells that are viable or dead based on their incorporation of different fluorescent dyes that stain live or dead cells. Potentially the individually sorted cell fractions could be sequenced separately. Currently this approach has been limited by the low yield of cells obtained after cell sorting from soil. However, newer platforms, including single-cell sequencing approaches, show promise for amplification of low DNA yields, and this could be a future direction for soil metagenomics.

Mining Soil Metagenomes

Soil represents a potential treasure trove for gene hunters because of the abundance of unknown genes that could potentially encode novel pharmaceuticals or other products of biotechnological interest. Two approaches are normally used for screening soil metagenomes for potentially interesting genes. The first is to rely on homology searches to gene databases. Using this approach, Hjort et al. identified chitinase genes in a metagenomic library from a phytopathogen suppressive soil. The other approach is to rely on screening of gene expression in clone libraries, a process that has been called "functional metagenomics". Functional metagenomics relies on expression of unknown genes of unknown origin in a foreign host. Since most microbes in soil have never been isolated and the majority of genes are unknown, this type of approach is ideally suited for screening soil for novel genes of interest. However, this approach is still hampered by several bottlenecks that result in very few hits when performing functional screens, including lack of efficient screens and low expression in heterologous hosts.

With the advent of shotgun metagenomic sequencing, the focus shifted from cloning into BAC and fosmid vectors to sequencing of total DNA. Depending on when soil metagenome projects were initiated, they were sequenced on different sequencing platforms with widely varying differences in read length, sequencing errors, and sequencing depth. The first shotgun soil metagenome of a Wisconsin farm soil was conducted using Sanger sequencing with long read lengths of high quality, but low depth. Therefore, only a fraction of the total community was sequenced, but still sufficient to distinguish key functional genes from other environments, such as whale fall and the Sargasso Sea. Similar broad functional comparisons between datasets were recently carried out on the Rothamsted park grass metagenomes, having higher sequencing depth, but shorter read lengths (using the 454 Titanium technology). Yergeau et al. used the 454 sequencing platform to compare an arctic soil active layer and permafrost, finding general differences in functional genes between the two soil types.

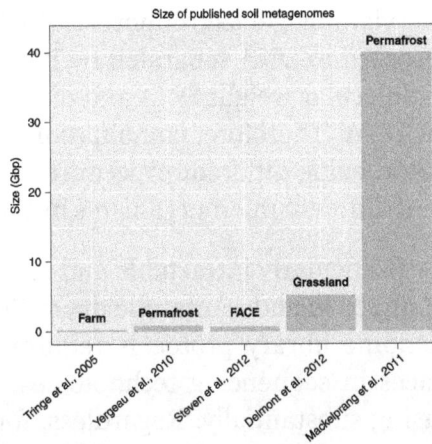

Progressively increasing sizes of soil metagenome datasets.

Sequencing with the Illumina GAII sequencing resulted in the highest amount of metagenome sequence data for soil. The Illumina sequencing reads were 2×113 bp in length, and 40 gigabases of sequence data were generated for 12 metagenomes (two active layer and two permafrost samples, before and after 2 and 7 days of thaw).

Current Challenges and Bottlenecks

There are several steps in the processing and analysis pipeline that are critical for soil metagenomics. The first step is DNA extraction. This is a challenge that was recognized in the 1980s and persists to this day. The problem is that there is no single method that has proven to have no bias. Also, since the true composition and diversity of any given soil have not been elucidated, there is no standard benchmark to determine which method is more accurate than any other. The cell lysis step in the extraction protocol is particularly problematic. Many soil bacteria are recalcitrant to enzymatic lysis procedures, such as actinobacteria. Therefore, depending on the lysis steps used, different members of the community may be more or less represented in the sequence data. In recognition of these biases, the Earth Microbiome Project recommends a standard lysis protocol be used for every sample contributed for metagenome sequencing analysis. Delmont et al. compared different commonly used soil DNA extraction protocols and found significantly different members of the soil community that were represented as a result.

Another option is to combine DNA extracts from several different extraction protocols to enable greater representation of the soil diversity. However, without knowledge of the true diversity and community composition of any given soil, it is currently not possible to know how representative the sequence data from a given DNA extraction method is.

Due to the complexity of shotgun metagenome sequence data from soils, there have been attempts to fractionate DNA extracts to reduce the complexity prior to sequencing. For example, Holben fractionated DNA based on GC content on density gradients. The different regions on the resulting gradient represented different microbial community compositions. Another way to fractionate is to take different sized DNA fragments that are separated by polyacrylamide gel lectrophoresis (PAGE). Delmont et al. found differences in genera represented in DNA separated by PAGE with higher amounts of some genera in the higher molecular weight DNA and different genera represented in the lower molecular weight DNA. Therefore, combination of different DNA fractions and/or DNA extracts prepared using different procedures is a promising approach to increase the representation of different microorganisms in a resulting soil metagenome.

Some soil environments are particularly intractable and result in very low DNA yields (nanogram quantities). Examples include Antarctic dry valley soils and arctic permafrost soils. Current metagenome library protocols require microgram quantities of DNA, although with advances in sequencing technologies, the amount required for DNA preparation is decreasing substantially. Regardless, for samples with extremely low DNA yields, it is still sometimes necessary to amplify the DNA to have enough for library construction. Two methods have been used to amplify DNA. One is the molecular displacement amplification approach (MDA) that has considerable bias. For example, the MDA approach was used to amplify DNA extracted from active layer and permafrost layer soil samples prior to shotgun metagenome sequencing on the 454 pyrosequencing platform. Another method recently used to amplify DNA prior to metagenome sequencing is emulsion PCR (emPCR). This method relies on amplification of individual size fractionated DNA segments in individual beads prepared in an oil emulsion. The DNA is diluted to the extent that no more than one DNA molecule should be present in an individual bead. Since each DNA molecule is independently amplified, there is no competition for primers from other DNA molecules. Recently, the emPCR technique was used to amplify low amounts of DNA extracted from permafrost soil.

Currently one of the biggest bottlenecks for soil metagenomics is the assembly step. The reason this is so challenging is because of the high microbial diversity in soil. The largest soil DNA metagenome sequence datasets to date are those for native prairie soils in the USA that are being sequenced as a Grand Challenge Pilot Study by the US Department of Energy Joint Genome Institute (JGI). For example, approximately 600 Gb of sequence data have been obtained for Kansas native prairie soil. However, even that amount of sequence is still far short of that required for efficient assembly.

There have been various estimates of the amount of sequencing that would be required to sufficiently cover a soil metagenome to enable reasonable assembly. Interestingly these estimates have been steadily increasing. After publication of the first shotgun metagenome sequence from soil, Tringe et al. estimated 2–5 Gigabases (Gb) would be required to get eightfold coverage of the most dominant members of the soil community. Delmont et al. estimated that about 450 Titanium pyrosequencing runs would be required to create contigs from all of the soil pyrosequence reads generated, but they recognized that chimeras might also be generated due to the complexity of the communities. One estimate based on an Alaskan soil metagenome was that 6 Tb of sequence data (or approximately 950,000 genome equivalents) would be required for the coverage of every OTU (at 97 % identity thresholds).

Depending on the soil type included in the metagenome sequencing, the assembly might be more or less problematic. For example, permafrost soils have a lower microbial diversity than most other soils and it was possible to use Velvet to assemble a draft genome of a novel methanogen from permafrost. The methanogen draft genome was represented in the majority of contigs greater than 1,000 Kb in length. It also appeared to have low sequence heterogeneity at the population level that aided in assembly. This case provides a clue that strong selection may enhance the ability to assemble specific members of the community that are selected by specific conditions and also may be present in higher relative amounts in the community.

Delmont et al. recently sequenced a grassland soil from the long-range experiment station in Rothamsted, UK, using 454 pyrosequencing to obtain nearly 5 Gb of sequence from 13 samples. Newbler assembly without optimization was used to assemble the reads into contigs. The largest contigs were about 23 Kb, but increasing the number of reads in the assembly did not improve the contig read lengths. However, the investigators noted that the contigs separated into two clusters represented by contig coverage. There was a trend for more sequences related to Firmicutes and Verrucomicrobia in the cluster of reads with $30\times$ coverage, whereas Proteobacteria were the majority of sequences in contigs with low coverage ($4.5\times$).

Annotation is also a current stumbling block for soil metagenomics due to the paucity of sequenced and annotated soil microbes in databases. Delmont et al. used the MG-RAST server (metagenomics.anl.gov) for annotation of the Rothamsted park grass soil metagenome. Out of 878 possible functional subsystems, 835 were found in the soil metagenome, thus illustrating the large potential functional diversity in the soil. The most abundant functional subsystems they found were related to cAMP signaling and Ton and Tol transport. These same subsystems were found in other existing soil metagenomes, including that of the Minnesota farm soil published earlier. The authors suggest that the high abundance of these subsystems in soil suggests that they have key role in soil ecosystems.

Soil Metagenome Examples

Some bacterial phyla are seen across several studies suggesting that they are prevalent soil microbes. These include members of the Actinobacteria, Chloroflexi, Fibrobacter, Acidobacteria, Planctomycetes, and Synergistetes that were found to be more prevalent in soil compared to the human gut or oceans. Mackelprang et al. also found similar phyla in active layer and permafrost soils from Alaska, and in addition sequences representative of the candidate phyla OD1 and OP11.

Yergeau et al. recently screened metagenomes from oil-contaminated arctic soils for hydrocarbon-degrading genes. They used BLAST to screen for major hydrocarbon degradation genes in the dataset. They found that contamination resulted in the increase of the abundance of several bacterial phylum/classes, including Actinobacteria, Alphaproteobacteria, and Gammaproteobacteria, and a decrease in the abundance of others, including Acidobacteria, Bacteroidetes, Chlorobi, Chloroflexi, Cyanobacteria, Firmicutes, Planctomycetes, and Deltaproteobacteria. Several of the classes that were enriched in the contaminated soils corresponded to well-known hydrocarbon degraders such as Pseudomonas, Rhodococcus, and Sphingomonads.

Steven et al. recently used shotgun metagenomics to compare soil microbial communities in biological soil crusts and creosote bush root zones that were collected at the Nevada FACE (free air carbon dioxide enrichment) site from soils with either ambient or elevated CO_2. They compared the phylogenetic representation from 16S rRNA gene pyrotag sequencing to classification of 16S reads from shotgun metagenomes. They found that the different approaches described the communities differently. They found a higher abundance of cyanobacteria in the biocrusts compared to the creosote root zones, but the proportion of cyanobacteria varied depending on the sequencing approach used. These authors cautioned against the use of different and incomplete databases for identification of taxonomic units as this may result in false ecological interpretation of the data.

Soil as a Habitat

A habitat is an area in which something lives. Soil is a habitat, and it is full of life.

The soil food web

The caterpillars of a native moth species burrow into soil. The vegetable caterpillar fungus uses the caterpillar for food. Its stem grows through the soil.

Around 25% of everything alive on the Earth uses soil as a habitat. Some animals live on top of the soil (in leaf litter or other organic matter), and others live below the surface. Some things live in the soil for their entire lives, and others live there for just a part of their lives. There are billions of microorganisms living in the soil too, but they are too small for us to see. Plants also live in soil. They depend on soil for air, water and nutrients.

Things living in the soil depend on each other and on non-living soil components like organic matter and minerals to survive. This interdependence and transfer of food energy is called a soil food web.

Soil Habitats Differ from Place to Place

Soil habitats are everywhere, even in cold dry places like Antarctica. Bacteria live in the soil. Soil habitats can be quite different. Soil characteristics affect what lives in it. For example, a soil that has lots of pore spaces for water and air usually supports more life than one consisting of hard clods. Temperature and rainfall also influence the types of plants and animals that live in the soil.

Soil Habitats in the City

The lawns and gardens around our houses and schools are home to all kinds of soil life, but we have to look to see it. Most soil organisms don't like to be exposed to sunlight. We even create soil habitats by building compost systems. A spadeful of soil and food scraps or grass clippings added to an empty bin soon becomes home to insects, earthworms and microscopic creatures.

It's just as well that soil organisms break down and recycle plant and animal wastes. Consider just one type of waste – insect exoskeletons. If these were never recycled, we'd be knee deep in insect body parts. Imagine all of the fallen leaves, potato peelings and cow dung that would still exist. Life above the ground wouldn't exist without the help of life below the ground.

Microbial Habitat in Soil

The microbial habitat is rarely studied in soil microbial ecology even though microbial cells are exposed and adapt to their local environmental conditions. The physical environment also constrains interactions among organisms. The nature of microbial communities and their functioning can only be fully understood if their habitat is accounted for.

In contrast to higher organisms, microbial cells are highly exposed to the conditions of their immediate environment and are rather limited in their ability to alter these conditions to suit their needs. In order to survive and grow, microorganisms adapt their physiology to the local physical and chemical conditions to which they are subjected. Drought and rewetting in soil, for example, induces wholesale changes in microbial physiology and has profound effects on the makeup of the microbial communities. The physical environment also constrains the range over which organisms can interact, thus affecting the boundaries, size, and composition of communities. In soil, microbial communities exist in a three-dimensional physical environment that is heterogeneous in both space and time and that regulates the flows of water, energy, gases and nutrients. It is likely, therefore, that microbial functioning and the nature of microbial communities can only be fully understood if the properties of their immediate environment and the manner in which they interact with their habitat are accounted for.

Soil microbial ecology is distinct from many other branches of ecology in that relatively little interest has been paid to the microbial habitat. A rapid analysis in Web of Knowledge TM shows that habitat is mentioned in approximately 10% of publications in soil microbial ecology, considerably less than the approx. 30% found in many branches of macro-ecology. This may be because the microbial habitat is not easily grasped, either from a measurement or a conceptual point of view, but there may also be a certain disregard among microbial ecologists for variations in environmental conditions and soil properties that occur at the scale of the soil aggregate or the microbial community. Nevertheless, the microscale variations in soil properties result in a myriad of micro-niches, which is believed to contribute to the high microbial diversity of soils from which soil biological functionality emerges.

The microbial habitat in soil can be viewed at different scales. At the coarsest scale, different soils constitute different microbial habitats. Biogeographic studies have shown that the properties of soils are significant drivers of microbial composition. However, the majority of the variability in bacterial diversity is not explained at these scales of analysis, but occurs at finer scales, suggesting that microbial interactions with their habitat at finer scales play an important role in structuring communities. Soils can be partitioned into different spheres of influence such as the rhizosphere, the detritusphere, or the drilosphere. The properties of these spheres of influence differentiate them from the bulk soil and have significant impacts on microbial communities. Microbial activity is significantly higher and the copiotrophic conditions result in a differentiation in microbial composition as well as a reduction in microbial diversity. The reduction in diversity may be due to the more amble supply of substrate, allowing the more competitive species to dominate. However, the physical structure of the rhizosphere is more spatially correlated which, coupled with the likely increase in substrate availability throughout the rhizosphere, may also result in a certain homogenization of the rhizosphere micro-niches, thus reducing microbial diversity. The same may be true of the detritusphere or the drilosphere.

Micro-habitat

Early observations of soil microbial communities in their natural environment suggested that they are predominantly found in pores with diameters between 0.8 and 10 μm and are associated with humified organic matter in the case of individual bacterial cells, or a diverse range of organic masses including plant and cellular remnants in the case of bacterial colonies. A number of technological developments (e.g., X-ray tomography, synchrotron-based spectroscopy or nanoSIMS) have allowed us to construct a more complete picture of the micro-environments in which microbial communities reside.

The structure of pore space in soil results in a huge diversity of micro-environments in which the biological component exists and is active. The diversity of micro-environments exists because the physical organization of solid and pore space causes a complex distribution of oxygen, water films and solute gradients, spanning distances as small as a few micrometers, to develop. These micro-environments are highly dynamic in nature as microbial activity, plant root growth or alterations in the water status can all affect their physical and chemical properties. During aggregate formation, for example, changes in the spatial organization of the solids and voids affect the architecture and connectivity of the pore space which can, in turn, affect the distribution of water films and the diffusion of substrate and gasses, the distribution of organic matter or the distribution of a range of elements. Lehmann et al. also showed that organic matter at the scale of the microbial habitat is highly heterogeneous in nature. This means that microbial communities residing in different micro-habitats experience different local environmental conditions, even when the overall environment of the soil is constant.

Relationship between Micro-habitat and Microbial Communities

The different local environmental conditions can be expected to exert different pressures on microbial communities, potentially affecting both the makeup of the communities and their activity. Indeed, the heterogeneity of the physical and chemical micro-environment in soil has been shown to have a significant structuring effect on microbial communities: microbial community structure is dependent on whether they are located within or at the surface of soil aggregates. Different regions of the soil pore space also harbor significantly different communities. The intimate relationship between microbial communities and their local habitat is further demonstrated by the rapidity with which community fingerprints respond to changes in aggregation. The phylogenetic composition of microbial communities also varies significantly among and within aggregates of the same soil. It should be noted that the differences among microbial communities from aggregates of the same soil and from different portions of the same aggregates were of the same order of magnitude as the differences measured in bulk samples of different soils, suggesting that an accurate picture of microbial communities cannot be obtained simply from bulk measurements.

Interaction distances or "calling distances" in the microbial world are extremely short. On leaf surfaces, for example, interactions among bacteria have been found to occur principally in the 5 to 20 µm range and diffusion limitations on solid surfaces reduce competition for a common resource to such an extent that a less competitive bacterial strain can coexist in close proximity ($\approx$ 100 µm) to a more competitive strain. In soil, the physical environment potentially constrains interactions among microbial populations even further, particularly in unsaturated conditions in which the aqueous zones in microbial habitats are fragmented. The few estimates of interaction distances in soil that have been made, suggest that the majority of microbial interactions occur over distances no greater than a few tens of micrometers. Although the numbers of microbial cells in soil are large, they only occupy a small part of the total surface area, a consequence of which is that individual microbial neighborhoods contain very few cells and even fewer species that are within interaction distances. If one considers a community to be a collection of individuals that interact with one another, then soil microbial communities may, in fact, be quite small and relatively species poor, which is in stark contrast to the very large numbers often used to describe soil microbial communities measured in bulk soil samples.

Microbial Communities

Communities can be defined as closed networks of interacting populations, in which the interactions among members of a community regulate members' relative abundances and determine the community's assembly rules. However, communities can also be "communities of a place"; in other words: a collection of individuals or species that exist in a defined space, generally because they tolerate the prevalent local conditions but do not necessarily interact or affect each other's existence. In view of the large disparity of scale between the size of the samples used for measuring microbial communities on bulk soil samples (0.25–0.5 g) and the distance over which microbial cells interact (100s µm), measurements of microbial community structure, composition or diversity cannot provide much reliable information on the identities of interacting species or the extent of the interactions. The measurements are more a reflection of "communities of a place", those microorganisms that can tolerate or thrive in at least one of the micro-habitats present in the soil. In these communities, habitat filtering regulates the composition of communities, as is often found for soil microbial communities rather than interactions among community members. One should therefore exercise caution when using co-occurrence network analysis for gaining insight into biotic interactions within soil microbial communities, as these analyses are unlikely to shed much light on microbial associations other than similar tolerances of soil environmental properties.

Microbial Communities and Activity

In communities in which there are causal interactions, the composition may affect eco-system functioning through competition for resources, facilitation, mutualism and a better coverage of habitat heterogeneity. Where there are no causal interactions, the composition of the communities is only likely to affect ecosystem functioning through a better coverage of habitat heterogeneity. In soil, more diverse microbial communities are more likely to be able to inhabit a greater portion of the available micro-habitats and to harbor a greater catabolic diversity. They might therefore be expected to better decompose the chemically heterogeneous soil organic matter. However, most studies suggest that there is little or no relationship between microbial community composition or diversity and soil organic matter decomposition in mineral soils. This is commonly ascribed to the existence of a high functional redundancy with respect to heterotrophic respiration in soil microbial communities. The functional redundancy theory requires that communities with low levels of diversity carry out the same functions across the available micro-habitats as more diverse communities. There are alternative explanations for the stability of a range of microbial functions to changes in composition or diversity of microbial communities. The first is that it reflects an averaging effect, as described by the law of large numbers. The law of large numbers states that the average obtained from a large number of realizations converges towards the expected value and, the greater the number of realizations, the closer to the expected value the average will be. If each microorganism–micro-habitat pair is viewed as a realization of a partic-ular activity, then even communities with low diversity levels (say with several tens or a few hundred taxa) would result in a large number of microorganism–micro-habitat pairs, the average activity of which would be expected to converge towards the average activity calculated from all possible combinations of microorganisms and micro-hab-itats for a soil. Although this explanation and the functional redundancy explanation are virtually indistinguishable from an experimental point of view, it has the merit of not requiring that different micro-organisms function in a similar manner. A second explanation is that microbial activity is constrained to such an extent by the environ-ment in which microbial communities exist, that all communities have similar activity, regardless of their composition or diversity.

Micro-habitat Constraints and Microbial Activity

The relationship between the availability of substrate and microbial respiration in soil is well established. Soils or regions of a soil, such as the rhizosphere, that have higher organic C contents generally have higher respiration rates. Interactions between the physical structure and water content of soil modulate respiration rates by regulating O_2 levels and the access that decomposing communities have to the substrate necessary for their activity. A number of studies have measured increases in respiration following the disruption of soil structure. The flush of respiration after physical disruption illustrates in a simple manner how fine-scale spatial disconnections between substrate and decomposers can constrain microbial activity. However, it is not yet clear what mechanism(s) regulate the availability of organic substrate to microbial decomposers. The significance of these regulatory mechanisms relative to biological or chemical limitations of activity has not been established either. Investigations into how the local environment constrains microbial activity have yielded contradictory results: no significant relationships between respiration in individual aggregates and aggregate properties that might be expected to influence the access microbial decomposers have to substrate (total pore volume or spatial proximity of organic matter and pores in tomographic images) have been found, and changes in the structure of the pore network do not appear to influence C mineralization at a given matric potential under steady-state conditions. Nonetheless, others have found that organic matter placed in different regions of the pore network is mineralized at different rates, suggesting that the local environmental conditions affect microbial activity. The differences, however, cannot be conclusively attributed to local environmental constraints because microbial communities also vary with the local environmental conditions, meaning that possible microbial community and local environment effects are confounded. In order to distinguish the roles of microbial communities and the local environment in regulating microbial activity, Nunan et al. used a reciprocal transplant approach in which microbial communities from two soils were placed in sterilized samples of either soil, in more or less connected aqueous environments. During the period of rapid growth, when the inoculated microbial communities were recolonizing the sterile environment using labile organic matter released from dead microbial cells, the respiratory rates of the different microbial communities were significantly different. As the incubation progressed, however, and the microbial communities decomposed organic matter unaffected by the sterilization process, the mineralization rates of all the communities decreased dramatically. The mineralization rates of the different microbial communities converged according to soil and connectivity of the micro-environment, regardless of the community of decomposers. This suggests a significant and dominant local environmental control on microbial decomposition of soil organic matter. The limitation may be such that intrinsic functional differences among communities cannot be expressed, resulting in a "habitat-induced functional redundancy".

The nature of the limitation is not clear, but in their influential study on N mineralization, found that there was a linear relationship between the activity and the square-root of time. They remarked that this would be expected for diffusion-controlled reactions. If C mineralization was constrained by the diffusion rates of organic substrates, then soil respiration curves would also be linear with the square-root of time. A re-analysis of some of our data shows that this can indeed be the case. If this was widespread, then a description of the pore and water structure of soil may be necessary to accurately describe C mineralization.

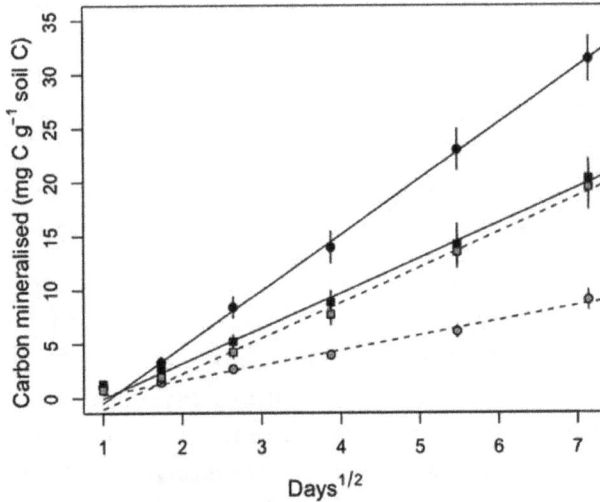

Linear relationship between cumulative C mineralisation and the square-root of time in topsoil (▪) and subsoil (•). Grey symbols are undisturbed samples and black symbols are sieved samples. Symbols show experimental data and lines show linear fit (R^2 all > 0.96). The measurements on the first day were not included in the linear fit these values may have been influenced by advection after adjustment of the soil moisture content. Bars show standard error of the mean, where error larger than size of symbol.

The heterogeneity of the soil habitat means that it may be more appropriate to consider soil as a juxtaposition of different "biomes" with different factors limiting process rates and microbial community assembly. In "biomes" in which the availability of substrate is relatively high and microbial communities can grow (i.e., the rhizosphere, the drilosphere or the detritusphere), the capacity of microbial communities to use the available substrate may dominate process rates and competitive exclusion may play a role in structuring microbial communities. However, where organic resources are less accessible, local environmental constraints, such as diffusion, may well regulate the rates at which processes occur. In these situations, interactions among micro-organisms are unlikely to structure the communities because resource availability is too low. Recent work on the soil microbial habitat has revealed a highly heterogeneous environment from both physical and chemical perspectives, which has significant influences on both the makeup and activity of microbial communities. This has still not been fully integrated into our understanding of the biological functioning of soil. It is likely to be profitable to pursue a line of research that better identifies the microbe-habitat interactions and the scales at which these interactions operate that structure microbial communities and regulate their activity.

Soil Biota

Soil biota is the biologically active powerhouse of soil which include an incredible diversity of organisms. Tons of soil biota, including micro-organisms (bacteria, fungi, and algae) and soil "animals" (protozoa, nematodes, mites, springtails, spiders, insects, and earthworms), can live in an acre of soil and are more diverse than the community of plants and animals above ground. Soil biota are concentrated in plant litter, the upper few inches of soil, and along roots. Soil organisms interact with one another, with plant roots, and with their environment, forming the soil food web.

As soil organisms consume organic matter and each other, nutrients and energy are exchanged through the food web and are made available to plants. Each soil organism plays a role in the decomposition of plant residue, dead roots, and animal remains. The larger soil organisms, such as millipedes and earthworms, shred dead leaves and residue, mix them with the soil, and make organic material more accessible to immobile bacteria. Earthworms can completely mix the top 6 inches of a humid grassland soil in 10 to 20 years. Ants and termites mix and tunnel through soils in areas of arid and semiarid rangeland. Predators in the soil food web include scorpions, centipedes, spiders, mites, some ants, insects, and beetles. They control the population of soil biota. The smaller organisms, including mites, springtails, nematodes, and one-celled protozoa, graze on bacteria and fungi. Other organisms feed on dead roots, shredded residue, and the fecal by-products of the larger organisms. The smallest soil organisms, microscopic bacteria and fungi, make up the bulk of the biota in the soil. They finish the process of decomposition by breaking down the remaining material and storing its energy and nutrients in their cells. Algae and fungi are the first organisms to colonize rock and form "new soil" by releasing substances that disintegrate rock.

Importance of Soil Biota

Through their interactions in the soil food web, the activities of soil biota link soil with the plants and animals above ground. Soil organisms perform essential functions that allow soil to resist degradation and provide benefits to all living things.

Residue decomposition: Without the soil food web, the remains of dead plants and animals would accumulate on the earth's surface, making nutrients unavailable to plants. Soil biota decompose these organic residues and some forms of organic matter in the soil. They convert these materials into new forms of organic matter and release carbon dioxide into the air. Many of the biota can break down pesticides and pollutants.

Nutrient storage and release: Most of the annual nutrient needs of rangeland plants are supplied through decomposition of organic matter in the soil. As soil organisms consume organic materials, they retain (immobilize) nutrients in their cells. This process prevents the loss of nutrients, such as nitrogen, from the root zone. When fungi and bacteria die or are eaten by other organisms, nutrients are mineralized, that is, slowly released to the soil in plant-available forms. Nutrient immobilization and mineralization occur continuously throughout the year.

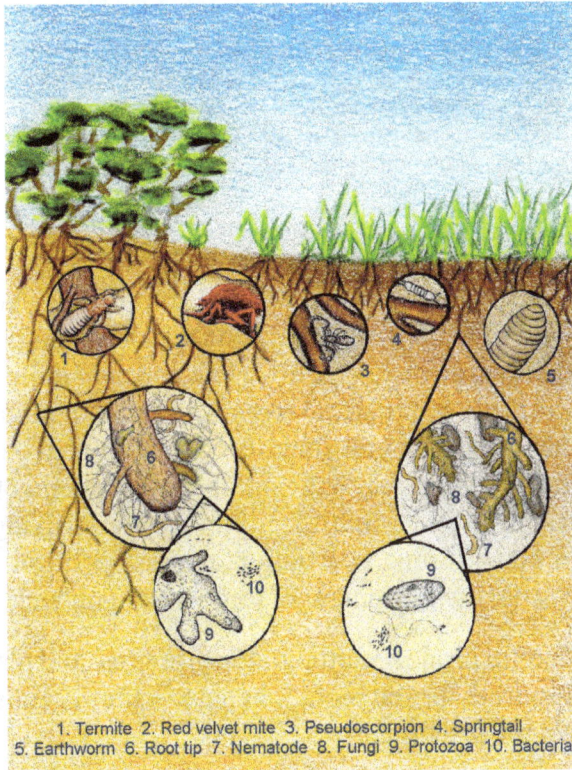

1. Termite 2. Red velvet mite 3. Pseudoscorpion 4. Springtail
5. Earthworm 6. Root tip 7. Nematode 8. Fungi 9. Protozoa 10. Bacteria

Some bacteria and fungi provide nutrients to plants in exchange for carbon. Special types of bacteria, called nitrogen fixers, infect the roots of clover and other legumes, forming visible nodules. The bacteria convert nitrogen from the air in the soil into a form that the plant host can use. When the leaves and roots die and decompose, nitrogen levels increase in the surrounding soil, improving the growth of other plants. Fungi produce hyphae that frequently look like fine white entangled threads in the soil. Some fungal hyphae (mycorrhizal fungi) attach to plant roots and act like an extended root system, providing nutrients and water to the plant.

Water storage, infiltration, and resistance to erosion: Soil biota form water-stable aggregates that store water and are more resistant to water erosion and wind erosion than individual soil particles. Threads of fungal hyphae bind soil particles together. Bacteria and algae excrete material that "glues" soil into aggregates. As they tunnel through the soil, the larger soil biota form channels and large pores between aggregates, increasing the water infiltration rate and reducing the runoff rate.

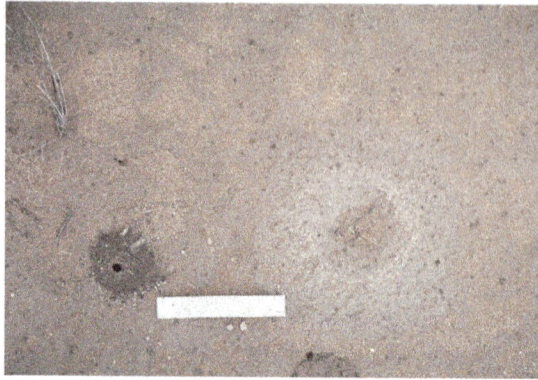

Left: Active ant mound. Right: Old ant mound.

Factors that affects Soil Biota

Soil biota multiply rapidly when organic material, roots, and plant litter, their food source, are available and the soil is moist and warm. Seasonal patterns of biological activity coincide with plant growth stages, litter fall, and root die-off. To be active, bacteria require films of water in soil pores, whereas fungi can function in drier conditions. When the soil is too dry, bacteria and fungi become less active or temporarily shut down, protozoa form dormant cysts, and the number of most other organisms declines. When the soil is saturated and anaerobic, the number of denitrifying bacteria increases. Organisms affect each other through predation and competition for food and space. Small soil pores can restrict the movement of large soil organisms.

Different types of vegetation produce different types of litter and plant residue and thus provide different food sources for soil biota. Changes in the vegetation or the pattern of plant distribution affect the soil organisms.

Management Considerations

Grazing: Proper management of the plant community is the best strategy for maintaining the benefits of the soil food web. Plant production and the supply of organic matter can be maintained or enhanced by timely grazing, the proper frequency of grazing, and control of the amount of vegetation removed. If the plant community is overgrazed, a reduction in the amount of surface plant material and roots will result in less food for soil organisms. As biological activity decreases, a downward spiral of the important functions of soil organisms results in a lower content of organic matter and impedes nutrient cycling, water infiltration, and water storage. Heavy grazing also can reduce the abundance of nitrogen-fixing plants, causing a decrease in the supply of nitrogen for the entire plant community.

Erosion: Erosion removes or redistributes the surface layer of the soil, the layer with the greatest concentration of soil organisms, organic matter, and plant nutrients. Runoff and wind erosion redistribute litter from one area of rangeland to a surrounding area.

The loss of organic matter reduces the activity of soil biota in the areas from which the litter has been removed.

Compaction by grazing animals and vehicles: Soil compaction reduces the larger pores and pathways, thus reducing the amount of habitat for nematodes and the larger soil organisms. Compaction can also cause the soil to become anaerobic, increasing losses of nitrogen to the atmosphere.

Fire and pest control: Fire can kill some soil organisms and reduce their food source while also increasing the availability of some nutrients. Pesticides that kill above-ground insects can also kill beneficial soil insects. Herbicides and foliar insecticides applied at recommended rates have a smaller impact on soil organisms. Fungicides and fumigants have a much greater impact on the soil organisms.

References

- What-is-soil, soils, soil-and-water, farm-management, agriculture: agriculture.vic.gov.au, Retrieved 21 April, 2019

- Physical-and-chemical-properties-of-soil, soil: biologydiscussion.com, Retrieved 15 January, 2019

- Soil-fertility-microbiology, soil-fertility, soil: biologydiscussion.com, Retrieved 21 July, 2019

- Soil-is-a-habitat: sciencelearn.org.nz, Retrieved 7 May, 2019

Understanding Soil Microbiology

2

Soil microbiology is a domain that deals with the study of microorganisms in soil. It also studies their functions and their effect on soil properties. Microbes play a vital role in making the nutrients and minerals in the soil available to plants. This chapter has been carefully written to provide an easy understanding of the varied facets of soil microbiology as well as the different kinds of microorganisms which are found in soil.

Soil microbiology is the scientific discipline that is concerned with the study of all biological aspects of the microorganisms (bacteria, archaea, viruses, fungi, parasites and protozoa) that exist in the soil environment. This is a subdiscipline of environmental microbiology.

Scope and Importance of Soil Microbiology

Living organisms both plant and animal types constitute an important component of soil. Though these organisms form only a fraction (less than one percent) of the total soil mass, but they play important role in supporting plant communities on the earth surface. While studying the scope and importance of soil microbiology, soil-plant-animal ecosystem as such must be taken into account. Therefore, the scope and importance of soil microbiology, can be understood in better way by studying aspects like:

1. Soil as a living system.
2. Soil microbes and plant growth.
3. Soil microorganisms and soil structure.
4. Organic matter decomposition.
5. Humus formation.
6. Biogeochemical cycling of elements.
7. Soil microorganisms as bio-control agents.
8. Soil microbes and seed germination.
9. Biological N_2 fixation.
10. Degradation of pesticides in soil.

Soil as a Living System

Soil inhabit diverse group of living organisms, both micro flora (fungi, bacteria, algae and actinomycetes) and micro-fauna (protozoa, nematodes, earthworms, moles, ants). The density of living organisms in soil is very high i.e. as much as billions/gm of soil, usually density of organisms is less in cultivated soil than uncultivated/virgin land and population decreases with soil acidity. Top soil, the surface layer contains greater number of micro-organisms because it is well supplied with Oxygen and nutrients. Lower layer/subsoil is depleted with Oxygen and nutrients hence it contains fewer organisms. Soil ecosystem comprises of organisms which are both, autotrophs (Algae, BOA) and heterotrophs (fungi, bacteria). Autotrophs use inorganic carbon from CO_2 and are "primary producers" of organic matter, whereas heterotrophs use organic carbon and are decomposers/consumers.

Soil Microbes and Plant Growth

Microorganisms being minute and microscopic, they are universally present in soil, water and air. Besides supporting the growth of various biological systems, soil and soil microbes serve as a best medium for plant growth. Soil fauna & flora convert complex organic nutrients into simpler inorganic forms which are readily absorbed by the plant for growth. Further, they produce variety of substances like IAA, gibberellins, antibiotics etc., which directly or indirectly promote the plant growth.

Soil microbes and Soil Structure

Soil structure is dependent on stable aggregates of soil particles-Soil organisms play important role in soil aggregation. Constituents of soil are viz. organic matter, polysaccharides, lignins and gums, synthesized by soil microbes plays important role in cementing/binding of soil particles. Further, cells and mycelial strands of fungi and actinomycetes, Vormicasts from earthworm is also found to play important role in soil aggregation. Different soil microorganisms, having soil aggregation/soil binding properties are graded in the order as fungi > actinomycetes > gum producing bacteria > yeasts.

Examples are: Fungi like Rhizopus, Mucor, Chaetomium, Fusarium, Cladasporium, Rhizoctonia, Aspergillus, Trichoderma and Bacteria like Azofobacler, Rhizobium Bacillus and Xanlhomonas.

Soil Microbes and Organic Matter Decomposition

The organic matter serves not only as a source of food for microorganisms but also supplies energy for the vital processes of metabolism that are characteristics of living beings. Microorganisms such as fungi, actinomycetes, bacteria, protozoa etc. and macro organisms such as earthworms, termites, insects etc. plays important role in the process of decomposition of organic matter and release of plant nutrients in soil. Thus, organic matter added to the soil is converted by oxidative decomposition to simpler nutrients/substances for plant growth and the residue is transformed into humus. Organic matter/substances include cellulose, lignins and proteins (in cell wall of plants), glycogen (animal tissues), proteins and fats (plants, animals). Cellulose is degraded by bacteria, especially those of genus Cytophaga and other genera (Bacillus, Pseudomonas, Cellulomonas, and Vibrio Achromobacter) and fungal genera (Aspergillus, Penicilliun, Trichoderma, Chactomium, Curvularia). Lignins and proteins are partially digested by fungi, protozoa and nematodes. Proteins are degraded to individual amino acids mainly by fungi, actinomycetes and Clostridium. Under unaerobic conditions of waterlogged soils, methane are main carbon containing product which is produced by the bacterial genera (strict anaerobes) Methanococcus, Methanobacterium and Methanosardna.

Soil Microbes and Humus Formation

Humus is the organic residue in the soil resulting from decomposition of plant and animal residues in soil, or it is the highly complex organic residual matter in soil which is not readily degraded by microorganism, or it is the soft brown/dark coloured amorphous substance composed of residual organic matter along with dead microorganisms.

Soil Microbes and Cycling of Elements

Life on earth is dependent on cycling of elements from their organic/elemental state to inorganic compounds, then to organic compounds and back to their elemental states. The biogeochemical process through which organic compounds are broken down to inorganic compounds or their constituent elements is known "Mineralization", or microbial conversion of complex organic compounds into simple inorganic compounds & their constituent elements is known as mineralization.

Soil microbes plays important role in the biochemical cycling of elements in the biosphere where the essential elements (C, P, S, N & Iron etc.) undergo chemical transformations. Through the process of mineralization organic carbon, nitrogen, phosphorus, Sulphur, Iron etc. are made available for reuse by plants.

Soil Microbes and Biological N_2 Fixation

Conversion of atmospheric nitrogen in to ammonia and nitrate by microorganisms is known as biological nitrogen fixation. Fixation of atmospheric nitrogen is essential because of the reasons:

1. Fixed nitrogen is lost through the process of nitrogen cycle through denitrification.

2. Demand for fixed nitrogen by the biosphere always exceeds its availability.

3. The amount of nitrogen fixed chemically and lightning process is very less (i.e. 0.5%) as compared to biologically fixed nitrogen.

4. Nitrogenous fertilizers contribute only 25% of the total world requirement while biological nitrogen fixation contributes about 60% of the earth's fixed nitrogen.

5. Manufacture of nitrogenous fertilizers by "Haber" process is costly and time consuming.

The numbers of soil microorganisms carry out the process of biological nitrogen fixation at normal atmospheric pressure (1 atmosphere) and temp (around 20 °C).

Two groups of microorganisms are involved in the process of BNF:

- Non-symbiotic (free living) and

- Symbiotic (Associative)

Non-symbiotic (free living): Depending upon the presence or absence of oxygen, non symbiotic N_2 fixation prokaryotic organisms may be aerobic heterotrophs (Azotobacter, Pseudomonas, Achromobacter) or aerobic autotrophs (Nostoc, Anabena, Calothrix, BGA) and anaerobic heterotrophs (Clostridium, Kelbsiella. Desulfovibrio) or anaerobic Autotrophs (Chlorobium, Chromnatium, Rhodospirillum, Meihanobacterium etc).

Symbiotic (Associative): The organisms involved are Rhizobium, Bratfyrhizobium in legumes (aerobic): Azospirillum (grasses), Actinonycetes frantic (with Casuarinas, Alder).

Soil microbes as Biocontrol Agents

Several ecofriendly bioformulations of microbial origin are used in agriculture for the effective management of plant diseases, insect pests, weeds etc. eg, Trichoderma sp and Gleocladium sp are used for biological control of seed and soil borne diseases. Fungal genera Entomophthora, Beauveria, Metarrhizium and protozoa Maltesia grandis. Malameba locustiae etc are used in the management of insect pests. Nuclear polyhydrosis virus (NPV) is used for the control of Heliothis/American boll worm. Bacteria like Bacillus thuringiensis, Pseudomonas are used in cotton against Angular leaf spot and boll worms.

Degradation of Pesticides in Soil by Microorganisms

Soil receives different toxic chemicals in various forms and causes adverse effects on beneficial soil micro flora/micro fauna, plants, animals and human beings. Various microbes present in soil act as the scavengers of these harmful chemicals in soil. The pesticides/chemicals reaching the soil are acted upon by several physical, chemical and biological forces exerted by microbes in the soil and they are degraded into non-toxic substances and thereby minimize the damage caused by the pesticides to the ecosystem. For example, bacterial genera like Pseudomonas, Clostridium, Bacillus, Thiobacillus, Achromobacter etc. and fungal genera like Trichoderma, Penicillium, Aspergillus, Rhizopus, and Fusarium are playing important role in the degradation of the toxic chemicals/pesticides in soil.

Biodegradation of Hydrocarbons

Natural hydrocarbons in soil like waxes, paraffin's, oils etc are degraded by fungi, bacteria and actinomycetes. E.g. ethane (C_2H_6) a paraffin hydrocarbon is metabolized and degraded by Mycobacteria, Nocardia, Streptomyces Pseudomonas, Flavobacterium and several fungi.

Soil Microbes

Organisms in the soil are both numerous and diverse. They range in size from the one-celled bacteria, algae, fungi, and protozoa, to the more complex nematodes and micro-arthropods, and to the larger organisms such as earthworms, insects, small vertebrates, and plants. Soil microbes (or microorganisms) are too small (i.e., smaller than 0. 1 mm) to be seen with the unaided eye. Bacteria are the most abundant microorganisms in soil, with a population of 10^{10}–10^{11} individuals and 6,000–50,000 species per gram of soil and a biomass of 40-500 grams per m^2.

If we are to understand microbial functions in soil and effects of management practices on soil quality, we need to consider more than just the number of individuals in a gram of soil. We also need analytical methods that will allow us to identify changes in the composition of the microbial community. Some of the more recently developed molecular genetic methodologies are proving useful in characterizing soil populations.

DNA sequencing is currently used to for taxonomic classification of microbes. Sequence information on nucleic acids (DNA- deoxyribonucleic acid and RNA – ribonucleic acid) associated with many microbial organisms is being generated rapidly. These sequences are analyzed by gene probe and polymerase chain reactions (PCR) technologies, which in turn allow us to detect organisms that previously could not be isolated or cultured.

Soil microbes play both beneficial (decomposition and nutrient cycling) and detrimental roles as pathogens and contributors to soil environmental problems such as global warming and groundwater contamination. The physical, chemical, and biological soil properties and their interactions with the resident community of soil microorganisms have a profound impact on growth and activity of microorganisms. As our understanding of these complex relationships develops, we should be able to develop soil management practices that are sustainable and that lead to maintenance and improvement of soil quality.

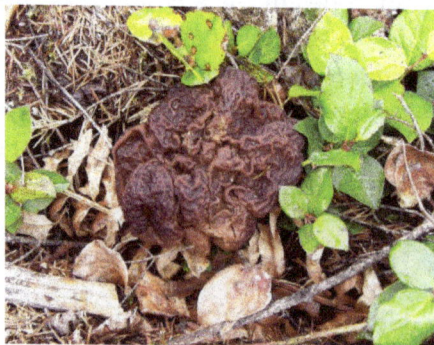

Brain fungus' Gyromitra esculenta at STEMS 3 site.

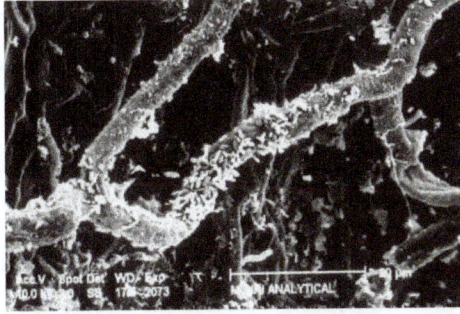

Scanning electron micrograph of mycorrhizal root tip hyphae coated with bacteria.

Microorganisms Found in Soil

Microorganisms which live in soil are algae, bacteria, actinomycetes, bacteriophages, protozoa, nematodes and fungi.

Soil Algae

Soil algae (both prokaryotes and eukaryotes) luxuriantly grow where adequate amount of moisture and light are present. They play a variety of roles in soil. One of the important role of blue-green algae is that it has revolutionised the field of agriculture microbiology due to use of cyanobacterial biofertilizer.

On the other hand they can also be used in reclamation of sodic soil i.e. alkaline soil, sewage treatment, etc. The prominent genera are Anabaena, Calothrix, Oscillatoria, Aulosira, Nostoc, Scytonema, Tolypothrix, etc.

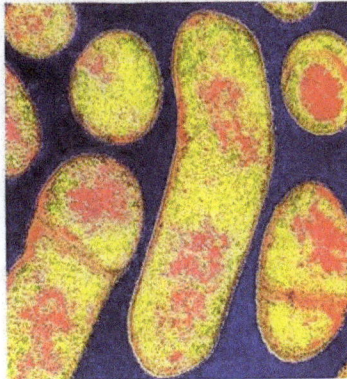

Clostridium botulinum, the anaerobic soil bacterium.

Bacteria

Bacteria are the smallest unicellular prokaryotes $(0.5 - 1 \times 1.0 - 2.0 \ \mu m)$, the most abundant group and usually more numerous than others, the number of which varies between 10^8 and 10^{10} cells per gram soil. However, in an agriculture field their number goes to about 3×10^9/g soil which accounts for about 3 tonnes wet weight per acre.

Based on regular presence, bacteria are divided into two groups:

a. Soil indigenous (i.e. true resident) or autochthonous, and

b. Soil invader or allochthonous.

Moreover, the number and types of bacteria are influenced by soil types and their microenvironment, organic matter, cultivation practices, etc. They are found in high number in cultivated than virgin land, maximum in rhizosphere and less in non-rhizosphere soil possibly due to aeration and nutrient availability.

The inner region of soil aggregates contained higher level of Gram-negative bacteria, while the outer region contained higher level of Gram-positive bacteria. This may be due to polymer formation, motility, surface changes, and life cycle of bacteria involved.

Bacteria do not occur freely in the soil solution but are closely attached to soil particles or embedded in organic matter; even after adding the dispersing agents bacteria are not completely dislodged from the soil particles and distributed in suspension as individual cell.

Moreover, they play a major role in organic matter decomposition, bio-trans- formation, biogas production, nitrogen fixation, etc. Example of some of soil bacteria is Agrobacterium, Arthrobacter, Bacillus, Alcaligens, Clostridium, Corynebacterium, Erwinia, Nitrosomonas, Nitrobacter, Pseudomonas, Rhizobium, Thiobacillus, etc.

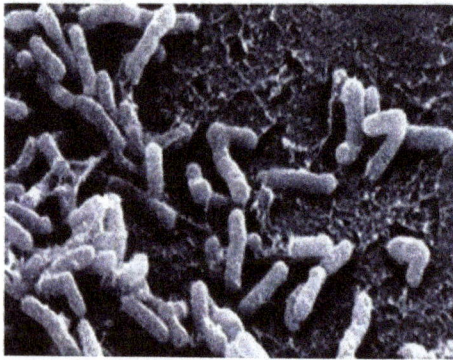

Electron micrograph of an Agrobacterium spp.

Actinomycetes

Actinomycetes share the characters of both bacteria and fungi, and they are commonly known as "ray-fungi" because of their close affinity with fungi. They are Gram-positive and release antibiotic substances. However, the earthy odour of newly wetted soils has been found to be a volatile growth product of actinomycetes.

Population of actinomycetes in soil remains greater in grass land and pasteur soil than in the cultivated land. In temperate zones the number of actinomycetes ranges from 10^5 to 10^8 per gram soil.

The most limiting factor is the pH which governs their abundance in soil. Its luxurient growth is favoured by neutral or alkaline pH (6.0 to 8.0). The important members of actinomycetes are: Actinomyces, Actinoplanes, Micromonospora, Microbispora, Nocardia, Streptomyces, Thermoactinomyces, etc.

Bacteriophages

Bacteriophages as well as plant and animal viruses have been observed in the soil. However, there role has not been clearly understood.

Protozoa

In moist soil most of the members of microfauna remain in encysted form. The population of each group is 103 per gram wet soil. The role of soil protozoa is predatory, as these eat upon bacteria and thereby regulate their population.

However, the number of protozoa can be correlated with plant root growth and indirectly with status of soil nutrients.

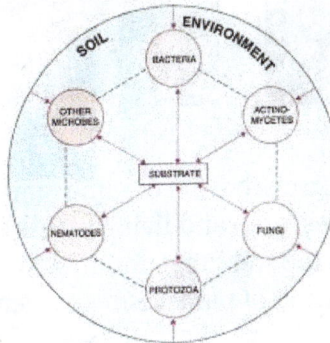

Microbial world of soil environment showing possible interactions between microbes and energy substrate, microbes and physical factors and between the microbes themselves (energy substrates include inorganic chemicals, dead organics and living plant root and soil fauna).

Nematodes

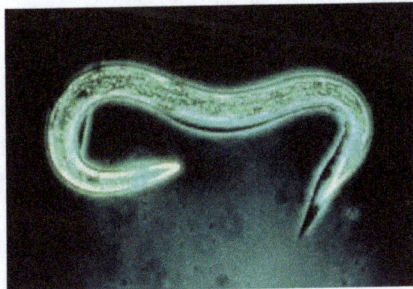

One square meter of forest soil teems with two to four million nematodes.

Until the role of nematodes in soil was understood nematology was in infancy stage. In recent years the ecology of nematodes has been greatly advanced. Nematodes derive nutrients for their growth and reproduction from the cell contents and cytoplasm of protozoa, bacteria, fungi, etc. Some common example of protozoa is Colpoda, Pleuro-tricha, Heteromita, Cercomonas, Oikomonas, Phalansterium, etc.

Fungi

In most of aerated or cultivated soils fungi share a major part of the total microbial biomass because of their large diameter and extensive net work of mycelium. However, population of soil fungi ranges from 2×10^4 to 1×10^6 propagules per gram dry soil and its number differs according to isolation procedure and composition of media.

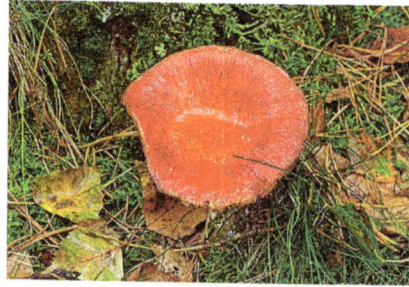

Representatives of two phyla of fungi: Ascomycetes and Basidiomycetes.

Fungi derive nutrients for their growth from organic matters, living animals (including protozoa, arthropods, nematodes, etc.) and living plants establishing different types of relationships. Garrett classified the soil fungi on the basis of substrate-specialization (i. e. fundamental niche) and duration of parasitism as given in figure.

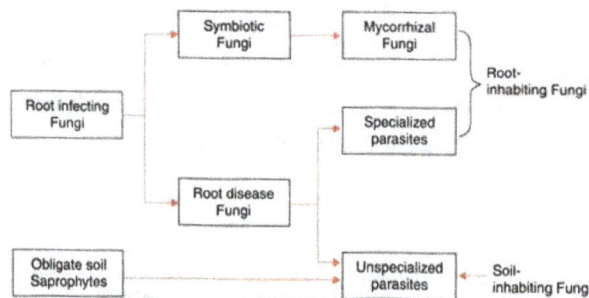

Classification of soil fungi.

The root inhabiting fungi are characterized by an expanding parasitic phase on the living host with a little declining saprophytic phase after the death of the invaded host. But the soil inhabiting fungi are characterized by the ability to survive indefinitely as soil saprophyte.

Success in competitive colonization of substrate by any particular soil fungus depends directly on its competitive saprophytic ability (CSA), inoculum potential at the surface of the substrate, and inversely as the aggregate inoculum potential of the competing fungi.

Many unspecialized root infecting fungi can probably exist as competitive soil sapro-phytes in the absence of living host, they have characteristics that confer a high degree of competitive saprophytic ability such as:

a. A high mycelial growth rate allied to rapid germination of resting propagules when stimulated by nutrient diffusion from a potential substrate,

b. A sufficient arrangement of tissue-degrading enzymes,

c. Production of fungistatic growth products including antibiotics,

d. Tolerance of those produced by other microorganisms.

However, most specialized parasites have evolved in such a way as to cause minimum possible damages to the host plant functioning as a piece of biological machinery exist-ing for support and protection of parasite.

The distribution of specialized parasite is determined by its host species. Some of the soil saprophytes are: Alternaria, Aspergillus, Cladosporium, Dematium, Gliocladium, Helminthosporium, Humicola, Metarrhizium, etc., and fungi associated with plant dis-ease are: Armillaria, Fusarium, Helminthosporium, Ophiobolus, Phytophthora, Plas-modiophora, Pythium, Rhizoctonia, Sclerotium, Thielaviopsis, Verticillium, etc.

Role of Soil Microorganisms

Although soil organisms comprise <1% of the total mass of a soil, they have a vital role in supporting all plants and thus animals. Some of their vital functions are described below.

Soil is Alive

Every gram of a typical healthy soil is home to several thousand different species of bacteria. One square metre of soil can contain about 10 million nematodes and 45000 microarthropods (springtails and mites). It has more species in it than 1 km² of rain-forest. In addition to bacteria, soil is home to microscopic fungi, algae, cyanobacteria, actinomycetes, protozoa and nematodes, and macroscopic earthworms, insects and the occasional wombat.

All of these organisms can be divided between autotrophs ("self-feeders"), such as plants, algae and cyanobacteria, and heterotrophs ("different-feeders"), such as fungi and bacteria, which decompose organic matter.

Soil Microbes Break Down Organic Matter

Microorganisms play an important role in the decomposition of organic matter. Different types of microbes are specialised to different types of organic matter, between them covering just about everything.

Soil Microbes Recycle Nutrients

Soil microbes play a crucial role in returning nutrients to their mineral forms, which plants can take up again. This process is known as mineralization.

Soil Microbes Create Humus

When the soil microbes have broken down all they can, what's left is called humus, a dark brown jelly-like substance that can remain unchanged in the soil for potentially millennia. Humus helps the soil retain moisture, and encourages the formation of soil structure. Humus molecules are covered in negatively charged sites that bind to positively charged ions (cations) of plant nutrients, thus forming an important component of a soil's cation exchange capacity. Humus is also suspected of suppressing plant diseases.

Soil Microbes Create Soil Structure

Some soil microbes secrete polysaccharides, gums and glycoproteins, which glue soil minerals together, forming the basis for soil structure. Fungal hyphae and plant roots further bind soil aggregates together. Soil structure is essential to good plant growth.

Soil Microbes Fix Nitrogen

Agriculture depends heavily on the ability of certain microbes (mainly bacteria) to convert atmospheric nitrogen (N_2 gas) to ammonia (NH_3). Some live freely in the soil, while others live in association with plant roots – the classic example is Rhizobiumbacteria in the roots of legumes. The process of conversion is known as nitrogen fixation.

Biological nitrogen fixation contributes about 60% of the nitrogen fixed on Earth. In contrast, manufactured fertilisers contribute 25%. As the cost of energy continues to rise, so too the cost of manufactured nitrogen fertilisers will rise, so biological nitrogen fixation is likely to have ever increasing importance in food production.

Soil Organisms Promote Plant Growth

Some soil microbes produce a variety of substances that promote plant growth, including auxins, gibberellins and antibiotics.

Soil Microbes Control Pests and Diseases

The best known example of the use of soil microbes in pest control is the commercial production of the soil bacterium Bacillus thuringiensis (Bt) to control caterpillar pests of crops. Some strains of Bt are used to control beetles and flies as well. Several strains of the fungal genus Trichodermahave been developed as biocontrol agents against fungal diseases of plants, mainly root diseases. Various other genera of fungi are used for the control of insect pests.

Importance of Soil Microbes

Living soil organisms have an important role in supporting all plants in their ecosystems. Soil itself can be viewed as a living organism and is home to thousands of different species of bacteria. One cubic foot can contain over 10,000 species of bacteria, 100 species of protozoa, 10,000 nematodes, 100s of species of algae and over 5000 individual insects. Not only does the soil contain beneficial bacteria but also contains microscopic fungi, algae, cyanobacteria as well as earthworms and other burrowing animals. These living beings amongst the soil are a key ingredient in keeping soil healthy and maintaining good soil structure (good drainage and gas exchange).

The buildup of organic matter in the soil profile over time can become a problem. Too much organic matter (primarily thatch) can start to reduce the efficacy of fertilizer applications, weed control as well as irrigation events. Thatch is also known to be a haven for unwanted pests and grubs. Healthy populations of soil microorganisms can help increase and speed up the breakdown of this organic matter and release helpful nutrients back into the soil in the form of nitrogen, phosphorous, potassium, hydrogen and various carbons (more on that shortly).

Speaking on conversion it is also important to note that a healthy population of soil microbes can help increase the speed at which applied fertilizer becomes available to the plant. Not all forms of each element are easily taken up by the plant. Soil microbes can help with the speed of which mineralization takes place and the nutrients are made available for plant uptake.

Earlier we touched a bit on carbon being added back to the soil when the microbes break down some of the organic matter in the soil. Not only is carbon one of the 17 essential nutrients to plant growth it is also one of the more versatile elements that can be added to the soil profile. Carbons are what promote the populations of soil microbes. One of the carbon sources that is created is humus (humic acid). Humic acid plays many roles in the soil profile, one of the roles is that it helps plants take up some of the more difficult elements for the plant to obtain (primarily metals), it does this through a natural chelation process. Humus can also help with moisture management on sandy soils, a good level of humus in the soil will help to retain soil moisture as well as improve the soil structure (drainage/pore space).

These are just some of the benefits a healthy soil microbial population can provide. In order to build a healthy population of these soil microbes you must provide them with an environment that favours their growth. Carbon/Organic based fertilizers are an excellent way to help provide them with an environment that they can thrive in.

Compost Microorganisms

In the process of composting, microorganisms break down organic matter and produce carbon dioxide, water, heat, and humus, the relatively stable organic end product. Under optimal conditions, composting proceeds through three phases: 1) the mesophilic, or moderate-temperature phase, which lasts for a couple of days, 2) the thermophilic, or high-temperature phase, which can last from a few days to several months, and finally, 3) a several-month cooling and maturation phase.

Compost Temperature

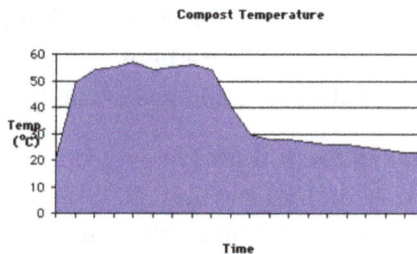

Different communities of microorganisms predominate during the various composting phases. Initial decomposition is carried out by mesophilic microorganisms, which rapidly break down the soluble, readily degradable compounds. The heat they produce causes the compost temperature to rapidly rise.

As the temperature rises above about 40 °C, the mesophilic microorganisms become less competitive and are replaced by others that are thermophilic, or heat-loving. At temperatures of 55 °C and above, many microorganisms that are human or plant pathogens are destroyed. Because temperatures over about 65 °C kill many forms of microbes and limit the rate of decomposition, compost managers use aeration and mixing to keep the temperature below this point.

During the thermophilic phase, high temperatures accelerate the breakdown of proteins, fats, and complex carboydrates like cellulose and hemicellulose, the major structural molecules in plants. As the supply of these high-energy compounds becomes exhausted, the compost temperature gradually decreases and mesophilic microorganisms once again take over for the final phase of "curing" or maturation of the remaining organic matter.

Bacteria

Bacteria are the smallest living organisms and the most numerous in compost; they make up 80 to 90% of the billions of microorganisms typically found in a gram of compost. Bacteria are responsible for most of the decomposition and heat generation in compost. They are the most nutritionally diverse group of compost organisms, using a broad range of enzymes to chemically break down a variety of organic materials.

Bacteria are single-celled and structured as either rod-shaped bacilli, sphere-shaped cocci or spiral-shaped spirilla. Many are motile, meaning that they have the ability to move under their own power. At the beginning of the composting process (0-40 °C), mesophilic bacteria predominate. Most of these are forms that can also be found in topsoil.

As the compost heats up above 40 °C, thermophilic bacteria take over. The microbial populations during this phase are dominated by members of the genus Bacillus. The diversity of bacilli species is fairly high at temperatures from 50-55 °C but decreases dramatically at 60 °C or above. When conditions become unfavorable, bacilli survive by forming endospores, thick-walled spores that are highly resistant to heat, cold, dryness, or lack of food. They are ubiquitous in nature and become active whenever environmental conditions are favorable.

At the highest compost temperatures, bacteria of the genus Thermus have been isolated. Composters sometimes wonder how microorganisms evolved in nature that can withstand the high temperatures found in active compost. Thermus bacteria were first found in hot springs in Yellowstone National Park and may have evolved there. Other places where thermophilic conditions exist in nature include deep sea thermal vents, manure droppings, and accumulations of decomposing vegetation that have the right conditions to heat up just as they would in a compost pile.

Once the compost cools down, mesophilic bacteria again predominate. The numbers and types of mesophilic microbes that recolonize compost as it matures depend on what spores and organisms are present in the compost as well as in the immediate environment. In general, the longer the curing or maturation phase, the more diverse the microbial community it supports.

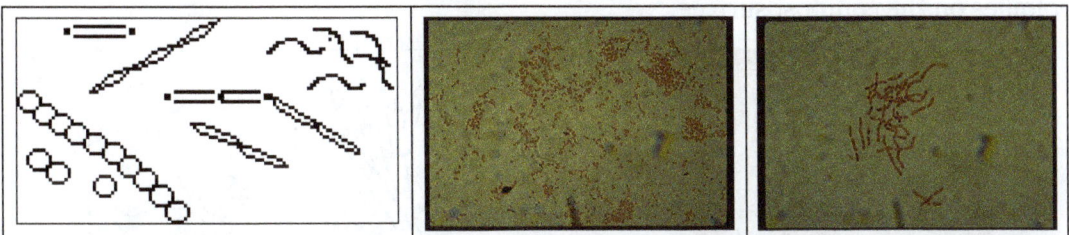

Actinomycetes

The characteristic earthy smell of soil is caused by actinomycetes, organisms that re-semble fungi but actually are filamentous bacteria. Like other bacteria, they lack nuclei, but they grow multicellular filaments like fungi. In composting they play an important role in degrading complex organics such as cellulose, lignin, chitin, and proteins. Their enzymes enable them to chemically break down tough debris such as woody stems, bark, or newspaper. Some species appear during the thermophilic phase, and others become important during the cooler curing phase, when only the most resistant com-pounds remain in the last stages of the formation of humus.

Actinomycetes form long, thread-like branched filaments that look like gray spider webs stretching through compost. These filaments are most commonly seen toward the end of the composting process, in the outer 10 to 15 centimeters of the pile. Sometimes they appear as circular colonies that gradually expand in diameter.

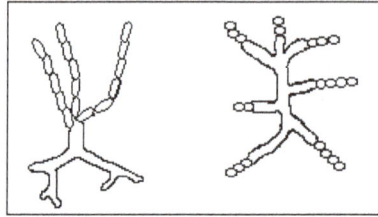

Fungi

Fungi include molds and yeasts, and collectively they are responsible for the decom-position of many complex plant polymers in soil and compost. In compost, fungi are important because they break down tough debris, enabling bacteria to continue the decomposition process once most of the cellulose has been exhausted. They spread and grow vigorously by producing many cells and filaments, and they can attack organic residues that are too dry, acidic, or low in nitrogen for bacterial decomposition.

Most fungi are classified as saprophytes because they live on dead or dying material and obtain energy by breaking down organic matter in dead plants and animals. Fungal species are numerous during both mesophilic and thermophilic phases of composting. Most fungi live in the outer layer of compost when temperatures are high. Compost molds are strict aerobes that grow both as unseen filaments and as gray or white fuzzy colonies on the compost surface.

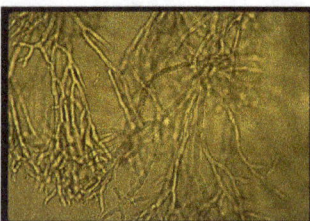

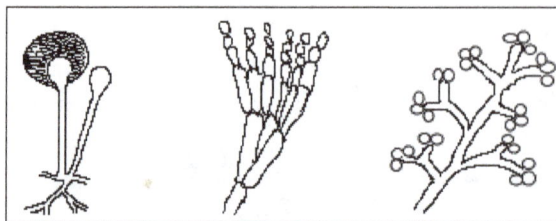

Protozoa

Protozoa are one-celled microscopic animals. They are found in water droplets in compost but play a relatively minor role in decomposition. Protozoa obtain their food from organic matter in the same way as bacteria do but also act as secondary consumers ingesting bacteria and fungi.

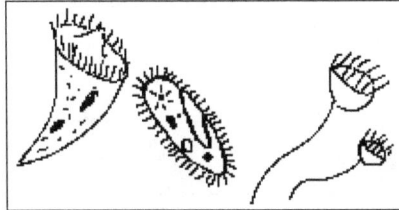

Rotifers

Rotifers are microscopic multicellular organisms also found in films of water in the compost. They feed on organic matter and also ingest bacteria and fungi.

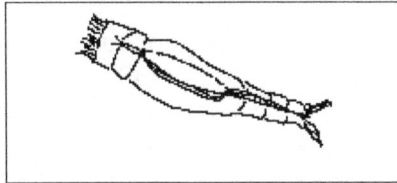

Microbial Biomass

Microbial biomass (bacteria and fungi) is a measure of the mass of the living component of soil organic matter. The microbial biomass decompose plant and animal residues and soil organic matter to release carbon dioxide and plant available nutrients. Farming systems that return plant residues (e.g. no-tillage) tend to increase the microbial biomass. Soil properties such as pH, clay, and the availability of organic carbon all influence the size of the microbial biomass.

The microbial biomass consists mostly of bacteria and fungi, which decompose crop residues and organic matter in soil. This process releases nutrients, such as nitrogen (N), into the soil that are available for plant uptake. About half the microbial biomass is located in the surface 10 cm of a soil profile and most of the nutrient release also occurs here. Generally, up to 5 % of the total organic carbon and N in soil is in the microbial biomass. When microorganisms die, these nutrients are released in forms that can be taken up by plants. The microbial biomass can be a significant source of N, and in Western Australia can hold 20 – 60 kg N/ha.

Microbial biomass is also an early indicator of changes in total soil organic carbon (C).

Unlike total organic C, microbial biomass C responds quickly to management changes. In a long term trial at Merredin, no significant change in organic C was detected between stubble burnt or retained plots after 17 years. Microbial biomass C in the same plots had increased from 100 to 150 kg-C/ha.

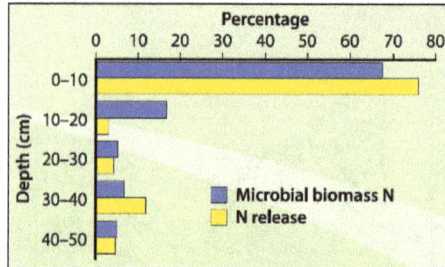

Microbial biomass nitrogen and release of nitrogen decreasing with depth.

In soil the microbial biomass is usually 'starved' because soil is too dry or doesn't have enough organic C. The amount of labile carbon is of particular importance as this provides a readily available carbon energy source for microbial decomposition. Soils with more labile C tend to have a higher microbial biomass.

Important sources of organic carbon as food for the microbial biomass are crop residues and soluble compounds released into the soil by roots (root exudates).

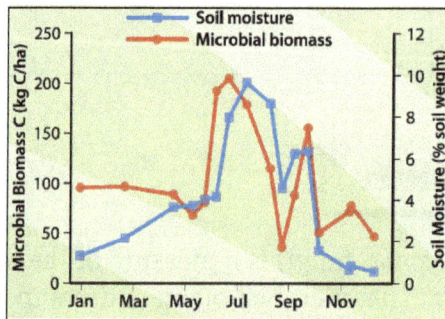

Microbial biomass carbon over a year from a soil near Meckering.

Factors affecting Microbial Biomass

The microbial biomass is affected by factors that change the water or carbon content of soil, and include soil type, climate and management practices. Rainfall is usually the limiting factor for microbial biomass in southern Australia. Soil properties that affect microbial biomass are clay, soil pH, and organic C. Soils with more clay generally have a higher microbial biomass as they retain more water and often contain more organic C. A soil pH near 7.0 is most suitable for the microbial biomass.

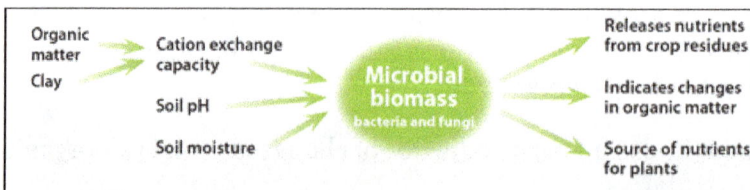

The main soil properties affecting the microbial biomass and factors influenced by it.

Management of crop residues influences microbial biomass as they are one of the primary forms of organic carbon and nutrients used by the microbial biomass. Retaining crop residues rather than burning them provides a practical means of increasing the microbial biomass in soil by increasing the amount of organic carbon available to them.

Soil Depth (cm)	Microbial biomass carbon (kg/ha)	
	Stubble retained	Stubble burnt
0 – 10	229	165
10 – 20	112	93
20 – 30	69	58

Tillage practices that are less disruptive to soil can increase the microbial biomass. Less disruptive tillage increases the microbial biomass by increasing labile carbon in soil. These management practices also protect soil aggregates and do not break fungal networks, which are an important habitat for the microbial biomass in soil.

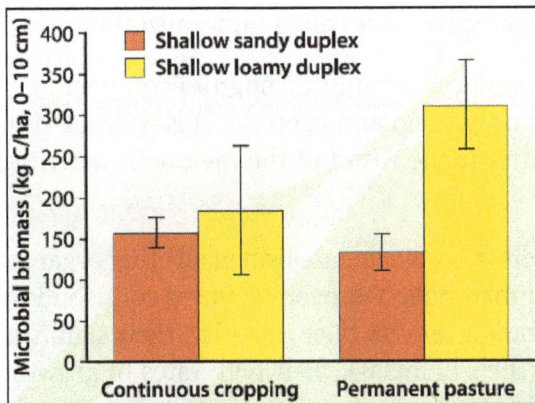

Microbial biomass in soils with different clay contents and under different management. Soils with more clay generally have a higher microbial biomass because they retain more water and often contain more organic carbon.

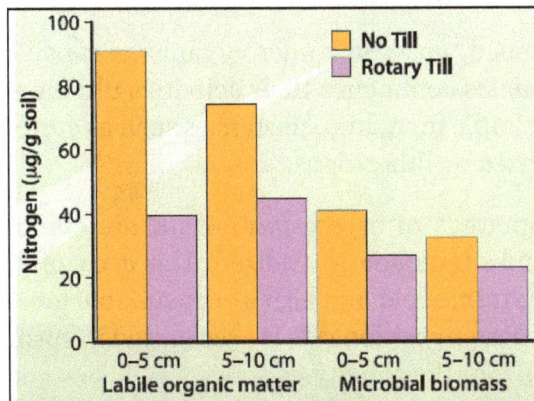

Increased nitrogen content in labile organic matter and the microbial biomass with no-till compared to rotary till in a 9-year field trial at Wongan Hills, WA.

The type of crops in a rotation can affect the microbial biomass. The residues of legume crops can increase microbial biomass due to their greater N contents. Rotations that have longer pasture phases increase microbial biomass because soil disturbance is reduced. This may not be the case in sandy soils, where the lack of clay means organic matter is broken down rapidly. This leaves the microbial biomass 'starved'.

Organic Matter and Microorganisms in Soil

The most important function of soil microorganisms is the decomposition of various kinds of organic matter present in the soil. Virtually all types of organic matters eventually find their way to the soil or to the sea. The soil organic matter chiefly consists of residues of dead plant and animals, and the excretory products of the living beings.

These organic constituents need to be converted into simple inorganic forms (minerals) to make them available to the autotrophic organisms. This conversion of organic matter into simple inorganic forms is called mineralization.

The mineralization is rendered mainly through decomposition of organic matter by soil microorganisms, mainly fungi and bacteria. It is estimated that 90% of the mineralization of organic matter is the result of the metabolic activities of these two groups of microorganisms.

The remaining 10% results from the metabolism of all other organisms, as well as the combustion of fuel and other materials. The overwhelming contribution of microorganisms to the process of mineralization reflects their ubiquity, their significant contribution to the bulk of living material (their biomass), their high rates of growth and metabolism, and their collective ability to degrade a vast variety of naturally occurring organic materials.

In general, the organic compounds of organism-residues are attacked by organotrophic microorganisms capable of digesting and oxidizing them.

As the oxygen is consumed, anaerobic microorganisms capable of reducing nitrates and sulphates or carbonates commence their activities ultimately creating an aerobic condition. The various simple inorganic substances such as ammonia, sulphide, hydrogen, etc. are avidly oxidized by lithotrophs.

Thus, the organic compounds of organism-residues are completely decomposed to simple organic forms and subsequently oxidized. This cycle of events occur where environment is suitable. Extreme cold and heat are unsuitable for normal turnover of the matter. In cooler parts, decomposition rate is accelerated in spring, whereas in warmer parts rainy conditions favour the activity.

The organic residues which are added to the soil can be categorized into three groups;

the easily degradable, moderately degradable and difficultly degradable. These three types of organic residues are attached by distinctly different types of microbes.

The result of microbial mineralization is the release of energy, water, gases, etc. on the one hand and formation of a complex amorphous substance (humus, process called humification) on the other hand. The whole process of mineralization terminating into humification is outlined in most simplified way in schematic representation of microbial decomposition of organic matter (mineralization) terminating to humification is outlined in most simplified way in figure.

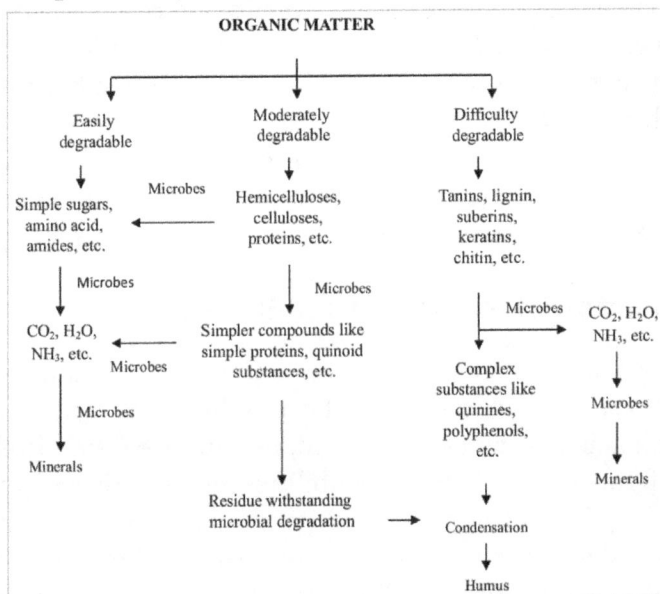

Schematic representation of microbial decomposition of organic matter terminating to humification.

The rate of oxidation of biologically important elements such as nitrogen, carbon, sulphur, phosphorus, etc. during mineralization of organic matter is of paramount importance to plants. This is so because these elements, though physically present in large quantities, may not be available to plants as they are not found in suitable chemical form.

These are the microorganisms which bring about these elements to the plants for their use through their activities and also make them free for reuse.

This process of 'use and reuse' through regular recycling of essential elements by the activity of microorganisms during the process of decomposition or mineralization of organic matter is called 'biogeochemical process' which is essential for the maintenance of life on our planet.

Thus, the organic matter added to the soil is converted by oxidative decomposition to simpler inorganic substances as a result of mineralization that are made available in stages for plant growth, and the residue transformed into humus as a result of the process called 'humification'; the processes of mineralization and humification running hand-in-hand in nature.

Role of Soil Microorganisms in Plant Mineral Nutrition

In their natural environment, plants are part of a rich ecosystem including numerous and diverse microorganisms in the soil. It has been long recognized that some of these microbes, such as mycorrhizal fungi or nitrogen fixing symbiotic bacteria, play important roles in plant performance by improving mineral nutrition. However, the full range of microbes associated with plants and their potential to replace synthetic agricultural inputs has only recently started to be uncovered. In the last few years, a great progress has been made in the knowledge on composition of rhizospheric microbiomes and their dynamics. There is clear evidence that plants shape microbiome structures, most probably by root exudates, and also that bacteria have developed various adaptations to thrive in the rhizospheric niche. The mechanisms of these interactions and the processes driving the alterations in microbiomes are, however, largely unknown.

The Interconnection of Plants with Soil Microbes

Although plant physiologists sometimes view soil as simply a source of nutrients to plants, it is actually a complex ecosystem hosting bacteria, fungi, protists, and animals. Plants exhibit a diverse array of interactions with these soil-dwelling organisms, which span the full range of ecological possibilities (competitive, exploitative, neutral, commensal, mutualistic). Throughout modern plant science, most interaction studies have focused on alleviating pathogenic effects such as herbivory and infection, or attenuating abiotic stress conditions. However, there has also been longstanding interest in characterizing the positive ecological interactions that promote plant growth. For instance, mycorrhizal fungi as well as the bacteria present in nodulated legumes were both recognized as root symbionts from the second half of 19th century. Already in the 1950s, crop seeds were coated with bacterial cultures (Azotobacter chroococcum or Bacillus megaterium) to improve growth and yield. By the 1980s many different bacterial strains, mainly Pseudomonas but also Azospirillum, had been described as having plant growth promoting effects. Since the 2000s, research focus has somewhat shifted away from individual microbial strains, and toward documenting the abundance and diversity of the root microbiome through metagenomics. Results from such sequencing studies have shown that the rhizospheric niche is a hotspot of ecological richness, with plant roots hosting an enormous array of microbial taxa. In the last few years, research has swung toward assembling rationally designed synthetic communities (SynComs) that comprise strains representing the dominant rhizospheric taxa, with the aim of re-capitulating favorable microbial functions under controlled experimental conditions. A major goal of this research field is to gain a mechanistic understanding of how soil microbes boost plant growth and defense, and then to use this knowledge to inform the optimal design of microbial communities tailored to carry out specific functions.

Microbial Traits and the Bioavailability of Nutrients for Plants

Three mechanisms are usually put forward to explain how microbial activity can boost plant growth:

(1) Manipulating the hormonal signaling of plants;

(2) Repelling or outcompeting pathogenic microbial strains;

(3) Increasing the bioavailability of soil-borne nutrients.

In natural ecosystems, most nutrients such as N, P, and S are bound in organic molecules and are therefore minimally bioavailable for plants. To access these nutrients, plants are dependent on the growth of soil microbes such as bacteria and fungi, which possess the metabolic machinery to depolymerize and mineralize organic forms of N, P, and S. The contents of these microbial cells are subsequently released, either through turnover and cell lysis, or via protozoic predation. This liberates inorganic N, P, and S forms into the soil, including ionic species such as ammonium, nitrate, phosphate, and sulfate that are the preferred nutrient forms for plants. In natural settings, these microbial nutrient transformations are key drivers of plant growth, and can sometimes be the rate-limiting step in ecosystem productivity.

Fertilization Practices and Environmental Sustainability

In most contemporary agricultural systems, macronutrients are provided through the application of mineral fertilizers. However, unsustainable fertilization practices are contributing to the large-scale alterations of Earth's biogeochemical cycles, through mechanisms such as soil degradation, waterway eutrophication, and greenhouse gas emissions. Furthermore, known reserves of phosphate rock are rapidly diminishing and predicted to be exhausted within a few decades, while N-fertilizer production via the energy-intensive Haber–Bosch process relies upon fossil fuels and thus exacerbates global warming and natural resource depletion. Due to the scale and severity of these fertilizer-induced problems, a current research priority is for agricultural science to develop alternative methods of sustaining plant nutrition with dramatically lower inputs of mineral fertilizers. One such possibility is to replace mineral fertilizers by organic inputs, and to supplement plants with specific root-associated microbes that depolymerize and mineralize the organic-bound nutrients. The logic of this idea is that organic inputs can be obtained more sustainably than mineral fertilizers, because myriad agricultural, industrial and municipal processes produce huge volumes of nutrient-rich "waste" that are currently disposed of, but could potentially be composted and applied as fertilizers. Another factor is that organically bound nutrients are more stable in the soil compared to mineral fertilizers, and therefore less prone to leaching and volatilization. Bio-fertilizers are already used in organic farming systems, but there is currently little mechanistic insight behind the choice of plant cultivars and microbial inoculants. This lack of precision occurs because of two major knowledge gaps: (1) it is unclear what strategies plants use to recruit beneficial microbes, and how much genetic variation exists for this trait; and (2) There is insufficient knowledge of which particular microbes are best partners for boosting plant nutrition from organic sources of N, P, and S.

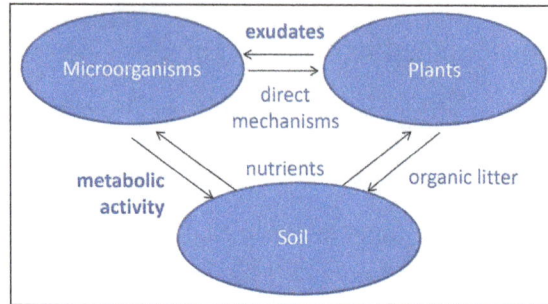

Interactions between plants, microbiota, and soil. Both plants and microorganisms obtain their nutrients from soil and change soil properties by organic litter deposition and metabolic activities, respectively. Microorganisms have a range of direct effects on plants through, e.g., manipulation of hormone signaling and protection against pathogens. Plants communicate with the microorganisms through metabolites exuded by the roots.

Effects of Plant Natural Variation on the Rhizosphere Microbiome

To selectively breed plants for optimized nutritional interactions with soil microbes, the genetic components of this trait must first be discovered. Sequence analyses showed differences between the composition of bacterial taxa in soil and plant rhizosphere or the endophytic fraction, showing that plants select for specific bacterial taxa and thus exert some control over their microbiomes. The next question is then to define the key genetic determinants that underpin how different plant genotypes interact with rhizospheric bacteria. Decades of research have shown that the susceptibility to pathogenic microorganisms is highly dependent on plant genome, between different species as well as within accessions of the same one. Similarly, Arabidopsis accessions showed large variation in supporting growth of rhizospheric bacterium Pseudomonas fluorescens in a hydroponic system. Indeed, sequence analyses have confirmed different microbiome structures across plant taxa, with bigger differences in more distant species and with also a larger contribution of environment and soil to the variation. When comparing accessions or varieties of the same species, genotypic effects on microbiome structure have been seen amongst Arabidopsis, maize, and barley. Regarding leaf microbiota, clear differences were shown for leaf microbiomes across 196 Arabidopsis accessions. The variation driven by plant-genome was particularly high for the most abundant operational taxonomic units (OTU). The variation was further explored by genome-wide association study (GWAS) using the number of reads for individual OTUs as quantitative phenotypes.

There, GWAS revealed that many of the significant SNPs linked to bacterial OTU structure were categorized as defense response, which was the most overrepresented gene ontology term among the candidate genes. In addition, genes involved in cell wall synthesis and kinase activity were enriched. Although several candidate genes affecting the leaf microbiome were identified, further confirmatory tests of mutants of these genes have not been reported and therefore the functionality of the genes in shaping the microbiome remains to be demonstrated. The leaf microbiome GWAS can be of major importance for understanding the processes in rhizosphere, because the leaf and root microbiomes are overlapping and might be shaped by similar processes. In an alternative approach to GWAS, Bodenhausen et al. monitored changes to SynComs inoculated onto leaves of Arabidopsis accessions and mutants of a priori selected genes. Again, a clear genotype effect upon microbial taxonomic composition has been observed in the 10 accessions and in several mutants. In three mutants, the effects were consistent and reproducible, two mutants were involved in cuticle synthesis and one in ethylene signaling (ein2). Given that only some 40 mutants and highly simplified SynComs were tested, the approach seems to be promising for analysis of root microbiome as well, particularly if mutants in nutrient uptake and assimilation would be investigated.

GWAS of Bacteria-Mediated Plant Traits

The sequence analyses, however, explore only the taxonomical composition of plant microbiome without taking into account the whole bacterial genomes or addressing the functions these microbes are performing. The best attempt so far to assess how plant genotype affects functional interaction with rhizobacteria is the analysis of variation in susceptibility of Arabidopsis accessions to the plant growth-promoting rhizobacterium Pseudomonas simiae WCS417r. The authors cultivated 302 accessions with and without the bacterium, which promotes changes in root architecture and growth through volatile emission. The accessions showed large difference in all three phenotypes scored: fresh weight gain, proliferation of lateral roots, and elongation of the primary root. Statistical GWAS analysis resulted in several highly significant associations, but despite some good correlation between the fresh weight and root architecture data, none of the positive SNPs were found with multiple phenotypes. The analysis led to identification of several candidate genes, but without further verification or confirmatory experiments. The analysis proves that GWAS of Arabidopsis accessions is a feasible approach to identify genetic loci that control the phenotypic variation in plant–microbe interactions. The challenge is to step beyond the relatively simple traits analyzed so far and to design screens that would allow to dissect the genetic architecture of the complex signaling and metabolic networks leading to variation in composition of root associated microbiota in different plant genotypes.

Plant natural variation in root exudates leading to distinctive microbial communities. We postulate that plant genotype has a strong effect on microbial community composition, mediated via root exudate composition. Consider three plant genotypes with differing root exudate profiles. At time 0 (upper panel), the three plants are transplanted from a sterile system onto a common soil where they are proximate to the same set of microbial strains. At time t (lower panel), the three plant genotypes have recruited distinctive microbial strains to their roots, which confer differential growth-promoting effects manifesting in different plant sizes.

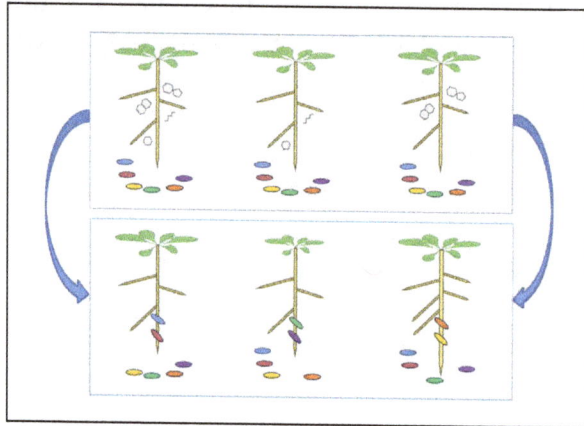

Plant Root Exudates

Metabolic Signals to Recruit Favorable Microbes

The growth of soil microbes is usually carbon-limited, so the high amounts of sugars, amino acids, and organic acids that plants deposit into the rhizosphere represent a valuable nutrition source. However, deposition of this labile carbon does not necessarily foster the recruitment of favorable microbes, because pathogenic strains can also use these molecules as growth substrates. Therefore, it can be postulated that plants have evolved recognition mechanisms to discriminate beneficial microorganisms from those that need to be repelled. In such a case, the specific molecules present in root exudates that contribute to shaping the microbial community structure are potential targets for plant breeding strategies that seek to engineer the rhizosphere microbiome. It has been shown that plant root exudates contain components used in belowground chemical communication strategies, such as flavonoids, strigolactones, or terpenoids. Studies on the microbiome of different plant species and accessions revealed strong variations, leading to the hypothesis that exudates are crucial in shaping plant–microbe interactions. Furthermore, it has been shown that plants specifically attract beneficial interaction partners via root derived signals.

Up to now, most information about signal perception and transduction in plant–microbe interactions comes from the field of plant pathology, where plant receptor-like kinases (RLKs) play a major role. In case of mutualistic interactions, nodulation and mycorrhizal interactions serve as model systems to identify recognition mechanisms between plants and microbes. In parallel to recognizing the microbial interaction partner by the plant, also microbes have to recognize their mutual interaction partner (the plant root). It is widely accepted that root exudates contribute to the establishment of the root microbiome. The term "root exudates" describes the molecules that are selectively secreted by roots and distinguishes it from the sloughing-off of root border cells. The overall release of fixed carbon compounds (border cells and exudates) into the surrounding soil is termed as rhizodeposition. Data about the amount of rhizodeposition range between 5 and 30% of the total amount of fixed carbon, which generally means a large loss of reduced-C for biomass and represents a considerable impact on the carbon budget of individual plants and also entire ecosystems. In a 14C approach, Hütsch et al. (2002) found remarkable differences in the amount of C-release among six different plant species ranging from 11.6 (wheat) to 27.7 (oil radish) mg C/g root dry matter. Also, the composition of these exudates varied between species, with oil radish exudates being rich in organic acids whereas pea exudates are rich in sugars. These data indicate that various plant species differentially modulate the chemical composition of their rhizospheres, which in turn might impact the associated microbial community. The recruitment of beneficial microbes might be crucial under environmental stress conditions such as nutrient limitation, pathogen attack, pests, high salt, or heavy metal stress.

Issues to Consider when Analyzing Root Exudates

To fully understand the dynamic interactions between soil microbes and plant roots, it is necessary to elucidate the specific molecules within root exudates that can recruit favorable microbial strains. This is a challenging problem in analytical biochemistry, because various biological and methodological issues must be addressed to undertake biologically insightful analyses of plant root exudates. Regarding cultivation, artificial plant growth systems cannot mirror the natural conditions in soil, but on the other hand, it is difficult to unravel the relevant communication signals occurring in soil, due to chemical interaction of metabolites with the soil matrix, and background metabolites released from decomposing organic matter or microbial exudation. Most analyses therefore settle on hydroponic cultivation, sometimes with an inert material to scaffold the roots. When sampling, the experimenter must choose whether to collect exudates in simple deionized water, or a more realistic medium containing mineral salts. Furthermore, it is effectively impossible to design an experimental approach that can differentiate exudates from sloughed-off border cells. A comprehensive summary on exudate collection and influences (e.g., pH, re-uptake by roots, incubation period) is presented in Vranova et al. For data acquisition, researchers are increasingly using unbiased mass spectrometry (MS) approaches such as gas chromatography (GC)-MS and liquid chromatography (LC)-MS.

However, detection of all metabolites in a sample is impossible due to physiochemical biases imposed by the selected extraction method, sample clean-up procedure, matrix effects and analytical technique. Therefore, different methods have to be combined for a comprehensive view on the metabolite profile. The subsequent analysis of the derived MS data is a huge challenge, beginning with data processing algorithms that enable feature detection, peak alignment and different normalization methods. These normalization and scaling algorithms have a large impact on the outcome of an analysis. To validate the identity of specific mass spectral features, fragmentation data (MS2 or MSn) are acquired and compared against publicly available databases, or authentic standards. Taken together, these challenges mean that comprehensive analysis of root exudates is not trivial.

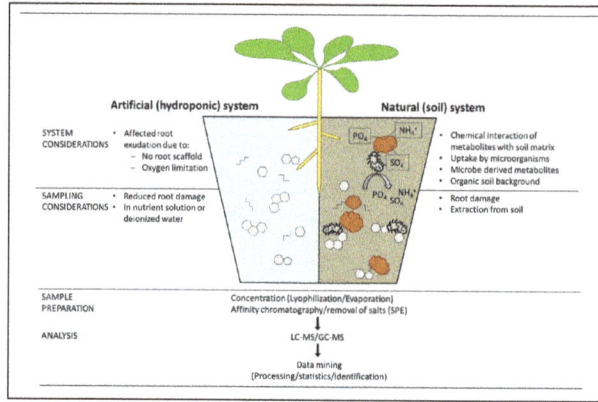

Technical considerations for analyses of root exudates. Comprehensive analyses of root exudate composition are crucial to advance our knowledge of plant–microbe interactions, but experimenters must consider the technical challenges at each step of the analytical workflow. First the growth system must be carefully considered, particularly whether to use soil or hydroponics, which each have advantages and disadvantages. At sampling, those using hydroponic systems must decide whether to collect the exudates in nutrient solution or deionized water, whereas those using soil systems must consider how to separate exudates from the soil matrix. During sample preparation, there are a multitude of options for sample concentration and clean-up, which will influence sample composition. Clearly, the mass spectrometry methodology used for data acquisition will play a crucial role in the workflow, because different setups allow the detection of different molecules. Finally, there is a growing awareness that data processing and analysis strategies also play a key role in shaping the derived data.

Recent Approaches to Analyze Root Exudate Composition

Various studies have described analyses of plant root exudates, with Phillips et al. developing a method to collect exudates from mature trees in the field, although microbial metabolism probably makes a significant impact upon this non-sterile system. That said, microbial nutrient uptake is an interesting aspect of plant–microbe nutritional interactions, with a fast degradation of flavonoid glucosides being observed by Carlsen et al. when comparing the flavonoid content in two soils and after different legume cultivations. To avoid microbial impact upon root exudate profiles, researchers have established diverse approaches of axenic hydroponic cultivation systems, which are easier to control, even though they represent artificial plant cultivation systems and plant responses might also include stress reactions due to oxygen limitation and insufficient root support. Furthermore, hydroponics is well suited for sampling of exudates, as the total liquid can directly been taken for further sample preparation procedures and root damage is minimized. However, collection of exudates widely ranges in timescale and the used collection medium (nutrient solution or water). Badri et al. collected root exudates from Arabidopsis thaliana in nutrient solution for 3 and 7 days for analysis by LC-MS and revealed that most compounds are present only after the longer incubation period. It could be hypothesized that this observation is due to sloughed-off border cells. Nevertheless, they also compared the exudate composition against root composition and stated an 80% difference based on detected molecular masses. Also Strehmel et al. applied a 7-day collection period in nutrient solution to obtain sufficient amounts of exudates from A. thaliana.

In contrast, Carvalhais et al. applied only a 6-h exudate collection period to Zea mays plants to minimize the effect of sloughed-off border cells, but used deionized water as collection medium. A similar approach has been used for barley root exudates that were collected for 4 h in deionized water. For a short-term exudate collection period from Arabidopsis a high amount of plants has been required to obtain sufficient amounts of exudates for LC-MS analysis. A direct comparison of different plant cultivation and exudate collection techniques revealed a huge impact on metabolite patterns. Especially, long incubations in deionized water may lead to overestimated exudation rates due to the high transmembrane gradient of solutes in low ionic strength medium. To date, most published data on exudates concentrate on specific metabolite classes such as primary metabolites, hormones, flavonoids, or phytosiderophores. Non-targeted metabolite profiling approaches of root exudates have been applied less frequently, although Strehmel et al. provided a comprehensive overview on secondary metabolites in Arabidopsis root exudates using LC-MS. In follow-up experiments, the data collection was complemented by GC-MS data and extended by a comparison of 19 Arabidopsis accessions, co-cultivation with Piriformospora indica and data from phosphate limitation. As MS technology continues to improve, it can be expected that more studies will undertake untargeted analyses of root exudate profiles.

How Root Exudates Differ Across Plant Genotype and Nutrient Limitation

If root exudate profiles are to be a potential breeding target for increasing plant–microbe nutritional cooperation, then it must be first understood how exudate composition varies across genotypes or in response to nutrient deprivation. Comparing exudate profiles across 19 natural Arabidopsis accessions showed a high natural variation for glycosylated and sulfated metabolites, such as flavonoids, glucosinolate degradation products, salicylic acid catabolites and polyamine derivatives. Regarding root exudation changes induced by nutrient limitation, it seems that phosphate deficiency results in a higher abundance of oligolignols and a lower abundance of coumarins. Future experiments could investigate a different panel of plant genotypes, perhaps where previous experiments have defined a phenotypic difference that is potentially linked to root exudation profiles. One possibility involves comparing genotypes that have shown contrasting affinity to recruit favorable microbes, or genotypes that differ in their nutrient starvation responses. Another option is root exudate profiling to analyze the phenotypic effects of mutants that were identified by GWAS studies.

Specific Molecules in Root Exudates

From our current knowledge of root exudates, is it possible to pinpoint a set of molecules that are particularly promising for recruiting favorable microbes to the rhizosphere? In legumes, it is well described that the flavonoid pathway has a huge impact on attracting rhizobia bacteria to roots and inducing NOD gene expression. Flavonoids are also crucial for hyphal branching and thus promoting mycorrhizal interaction. Both of these interactions result in increased plant nutrient uptake, with mycorrhiza and root nodules boosting phosphorus and nitrogen, respectively. Perhaps other plants and soil bacteria have also tapped into this signaling pathway, and further analysis of root exudates will give us clues as to whether flavonoids play a role in this communication leading to increased nutrient uptake. From analyzing plant mutants, candidate genes that have been investigated so far are related to the transfer of metabolites into the rhizosphere, hormonal signaling, or to biosynthesis, e.g., genes from phenylpropanoid pathway. These results give us some hints about candidate molecules that are exuded by plant roots to recruit beneficial bacteria, such as strigolactones and flavonols, and presumably further analyses will expand this list. To fully exploit this knowledge, then it is desirable to identify which microbial strains are recruited by these molecules, and what benefits they confer to the plant.

Using Sequence Data to Predict Microbial effects on Plant N, P, and S Nutrition

Mechanisms of Microbial Nutrient Provision to Plants

Plants give large amount of carbon away to the rhizosphere that nourishes soil microorganisms. So what do plants get back? In natural soils the vast majority of N, P, and S atoms are organically bound, while in the atmosphere the vast majority of N is contained in the N_2 molecule. Due to the different metabolic capacities of plants and microbes, these nutrient sources are minimally bioavailable to plants, but can be metabolized by various soil microbes. This means that nitrogen fixing and nutrient mineralization processes carried out by soil microbes are crucial for plant nutrition in natural ecosystems, because these reactions metabolize recalcitrant forms of N, P, and S to liberate these elements for plant nutrition. It must be briefly stated that this established paradigm has been somewhat questioned in recent years, as several studies have demonstrated direct plant uptake of various organic-N forms. However, it is still generally accepted that microbes are better competitors for these nutrients due to the low diffusivity of organic-N molecules in soil, and the results of isotope labeling studies generally support the concept that most organic-N is first assimilated by microbial taxa, then subsequently assimilated by plants upon microbial turnover.

Over several decades, the soil microbiology literature has accumulated a list of microbial metabolic processes that are linked to plant N, P, and S nutrition. Commercially, the symbiotic association between legumes and bacteria is routinely exploited when field crops are inoculated with nitrogen-fixing rhizobia strains. But, how can this phenomenon be refined and optimized for widespread use in more sustainable agricultural systems—not only for nitrogen fixation in legumes, but also for N, P, and S nutrition in non-leguminous crops? Given that different bacterial strains exhibit differing metabolic capabilities, coupled with the huge amount of genomic sequence data from soil microbes that has recently been generated, one possibility is to pinpoint the genes that encode metabolic pathways for agriculturally beneficial N, P, and S metabolism, and to boost the microbes that contain these specific genes in agricultural soils.

Cultivation-Independent or Cultivation-Dependent Approaches

The major difficulty in investigating soil microbial communities is that only a small fraction of the inhabiting taxa can be cultivated in the laboratory, where experimenters can undertake detailed and controlled analyses. Therefore, the literature investigating root-associated microbes can be roughly divided into either cultivation-independent or cultivation-dependent studies. Generally, most cultivation-independent approaches extract root-associated microbes in situ, and then analyze the properties of this community. In contrast, cultivation-dependent approaches generally inoculate soil or root-associated microbes onto laboratory growth medium, before analyzing distinct strains that have been cultivated in the laboratory. For cultivation-independent approaches, techniques typically used to analyze soil microbiota often include: (1) 16S sequencing and PLFA measurements to infer taxonomic breakdown; (2) metagenomic, metatranscriptomic, or metaproteomic analyses to infer functional capacity of microbial communities; and (3) enzyme assays, respiratory measurements, or substrate utilization assays to measure functional activity of microbial communities. In cultivation-dependent approaches, the analytical possibilities are multitude—once root-associated organisms have been cultivated in a laboratory setting, they can be analyzed with any available technique. In the literature, there are many studies that use both cultivation-dependent and cultivation-independent techniques to draw links between microbial properties and plant nutrition.

Key microbial metabolic processes related to plant nutrition.

Element	Biochemical	Microbial
	process	genes
Nitrogen	Nitrogen fixation	nifD, nifH, nifK
	Protein depolymerization	apr, npr, sub
	Urea catabolism	ureA, ureB, ureC

Phosphorous	Phosphate ester cleavage	phoA, phoD, phoX, ACPase, glpQ, ushA, appA, phyA, phyB
	Phosphonate breakdown	phnJ, phnX
Sulfur	Sulfate ester cleavage	aslA, asfA
	Sulfonate breakdown	ssuD

Soil microbial metabolism boosts plant nutrition by converting recalcitrant forms of N, P, and S to forms that are more bioavailable for plant uptake. This table collates a set of well-known microbial metabolic processes that contribute to plant N, P, and S nutrition.

Microbial Taxa and Metabolic Pathways

Several studies have used cultivation-independent approaches to investigate microbial community structure/function across soils exposed to different fertilization regimes. Typically, these studies compare soil microbiota from highly fertilized soils against those from unfertilized or poorly fertilized soils, in situations where the low-fertilization regimes have encouraged mutualistic nutrient transfers between plants and microbes. From analyses of microbial taxonomy, it has been shown that the abundance of certain bacterial taxa is related to amount of applied fertilizer, with the copiotrophic phylum Actinobacteria being positively correlated with N fertilization, whereas the oligotrophic phylum Acidobacteria is negatively correlated. However, the results of a meta-analysis suggest that it is difficult to generalize a consistent response of microbial taxon abundance to N fertilization, because local environment and management play a dominant role in shaping microbial community structure. The work of Hartmann et al. examines bacterial 16S and fungal ITS2 sequences in a long-term field experiment comparing organic with conventional farming systems, showing correlations between taxon abundance and fertilization regime, with the bacterial Firmicutes phylum and several fungal taxa being more abundant on soils fertilized with manure.

The last 15 years has seen an explosion in the number of rhizospheric microbiome sequencing studies, offering new taxonomical insights into the microbial communities associated with plants. However, the utility of taxonomic analyses for predicting microbial community function can be questioned, because in the bacterial literature it is becoming increasingly apparent that taxonomic groupings derived from 16S homology are imperfect predictors of a bacterial strain's functionality. Recent studies have sequenced the whole genomes of several closely related strains, and have discovered that although these strains are categorized as closely related due to the presence of a homologous core genome, in fact there can be considerable divergence in the accessory genome, meaning that the encoded functional capacities will also be significantly different. By sequencing the bacterial 16S gene, researchers infer the phylogeny of a strain's core genome, but it can be argued that this gives little information about metabolic traits, because many of the key genes involved in N, P, and S metabolism are accessory genes, which are not taxonomically conserved due to the high prevalence of horizontal gene transfer between bacteria. Therefore, metagenomics studies that profile the abundance of all gene sequences (not just 16S) should have more power to unravel links between microbial genetics and plant nutrition, although it should be remembered that only a small proportion of the soil DNA pool is actively expressed. From metagenomic studies that compare the effects of different fertilizer inputs, it seems apparent that certain genes are more abundant in soils with lower fertilizer inputs, such as urea metabolism and unclassified metabolic genes. These genes are thus positioned as potential targets for improving the microbial provision of plant-bioavailable N, P, and S. However, the sheer complexity of the soil microbiome makes it difficult to draw mechanistic links between specific genes and ecosystem processes, which is one of the reasons why many researchers are adopting SynCom experiments that attempt to re-construct a simplified rhizosphere microbiome in a controlled setting.

In the soil biology literature, enzyme assays have established a set of enzymes linked to high-functioning soil microbiota, such as protease, urease, various phosphatases, and sulfatase. Therefore, the bacterial strains that possess the genes encoding these proteins are candidates for boosting nutrient transfers to plants. However, one challenge involves managing the stoichiometric availability of different nutrients to promote the activity of these enzymes. In the priming literature, it is generally accepted that soil microbiota are usually limited by the amount of labile C. Supply of labile carbon (e.g., root exudates) can relieve this limitation, such that N, P, or S then becomes the limiting nutrient, and microbes then express enzymes that can depolymerize recalcitrant forms of these nutrients. So, even if the soil microbiota contains strains with genes encoding the aforementioned enzymes linked to soil health, soil conditions must be optimized for these microbial proteins to be expressed and active. Another method to measure metabolic capacity of soils involves community level physiological profiling assays, which measures substrate degradation affinity across different fertilization regimes, although usually these assays are designed with an emphasis on degradation of C-sources rather than sources of N, P, and S. This technique has been applied to soil receiving different fertilization practices, and it has been shown that the capacity to degrade a diverse range of substrates is correlated to other aspects of soil health, such as organic carbon content and disease suppression. Perhaps future studies could modify this approach to develop assays that measure the capacity of soil microbes to degrade various sources of N, P, and S. However, the mechanistic insight derived from these assays is sometimes questioned, because they measure the capacity of soil microbiota to grow on specific nutrients under laboratory conditions. This setup may therefore select for a small number of fast-growing taxa and not correlate with the in situ activity of these substrate degradation pathways.

Microbial Genes in Cultivation

It can be posited that soils with high rates of microbial N, P, and S cycling should harbor microbes with specialized genes encoding these traits, and therefore that microbial strains isolated from these soils should possess useful metabolic attributes for boosting plant nutrition. Interestingly, Rhizobia strains isolated from N-fertilized soils exhibited a lower capacity to promote plant growth compared to strains isolated from adjacent unfertilized plots, indicating that the management history of the isolation site impacts the degree of mutualism in the resulting isolate. In the P literature, a similar phenomenon has been observed, with phosphate mineralization being more common in isolates from soils where bioavailable-P was less abundant. For sulfur, one example is from soils from the Rothamsted Broadbalk experiment, where different S-fertilization practices had led to fields that exhibit high versus low sulfatase activity. Bacterial strains were isolated from these contrasting soils, and functional assays such as enzyme measurements and growth on minimal media revealed that the strains isolated from low-SO_4^{2-} soils contained several mechanisms for depolymerizing organic-S. Together, these results imply that research programs seeking "elite" microbial strains that can maximally boost plant nutrition could begin with inocula from sites that favor plant–soil feedbacks, such as unfertilized soils or organic farms.

Microbial Strains that Promote Plant Growth by Enhancing N, P, and S Nutrition

To truly be useful in an agricultural setting, it must be proven that candidate growth-promoting strains can be re-inoculated onto plants, successfully colonize the rhizospheric niche, and then mediate nutrient mobilization that benefits plant growth. This can be tested through plant–microbe interaction assays, where candidate strains are tested for their ability to promote plant growth and nutrient acquisition. Once again, this research field is most mature for the case of nitrogen-fixing Rhizobia, where decades of research have endeavored to define the optimal inoculation practice, searching for the right combination of plant genotypes and rhizobia strains to suit specific climates and soils. Regarding the taxonomy of nitrogen-fixing symbioses, it should be mentioned that nitrogenase genes are present in diverse bacterial taxa, and that non-leguminous plants have been documented to host N_2-fixing bacterial strains, perhaps implying that other plant–microbe combinations (not just legumes and Rhizobia) could be similarly optimized to promote nitrogen fixation. There are also reports of plant growth promotion via microbial mobilization of other nitrogen sources, shown by higher yield in plants inoculated with bacterial strains, although for one of these experimental setups, it seems that the source of this N was directly from the ammonium sulfate fertilizer rather than from organically bound soil N. It has been shown that unsterilized grass seeds can better access protein-N compared to sterilized seeds, but the specific strains that provide this service were not elucidated. The ability of the fungi Glomus intraradices to transfer organic nitrogen to plants has also been shown, suggesting that future experiments could focus on documenting other fungal strains with this capacity and characterizing the relevant genes and mechanisms. For phosphorous, the literature contains a large number of reports of both fungal and bacterial strains with the capacity to solubilize inorganic P, and also many reports of strains that can mineralize organic P. Many of these P-mobilizing strains are also characterized as growth promoting microbes, but microbial promotion of plant growth can operate through a wide variety of mechanisms, and sometimes it is not conclusive that P-mobilization is responsible for the plant growth promotion elicited by these strains. For sulfur, studies of a plant growth-promoting Pseudomonas strain have used genetic knockout of the sulfonate monooxygenase enzyme to show that organic-S mineralization accounts for some fraction of the growth-promoting phenotype.

The research field is beginning to build large collections of genomically sequenced bacterial isolates that can be re-assembled into SynComs. Xia et al. isolated endophytic bacterial strains from plants grown under organic management, and showed that over half of these strains can boost tomato growth in a greenhouse experiment. This high proportion of growth-promoting isolates shows that the capacity to promote plant growth is widespread amongst plant-associated bacteria, but to gain a mechanistic understanding into how these growth-promoting effects are manifested, it will be necessary to conduct detailed investigations into the genetics, biochemistry and physiology of these growth-promoting strains. Furthermore, microbial community experiments should also consider how the interaction between different strains affects plant growth promotion. Such knowledge will enable the rational selection of growth-promoting strains and communities, driven by defined genetic and biochemical mechanisms.

A Growing Community: Challenges and Perspectives in Modeling Plant–Microbe Interactions

With the vast quantity of data being generated by high-throughput experimental techniques, new opportunities are arising to integrate theoretical and computational approaches with experiments. Potential synergies include designing hypothesis-driven experiments based on the results of modeled scenarios, or using modeling as a tool to mechanistically interpret the results of high-throughput experiments. In the bacterial field, there is currently a major initiative to integrate genomic sequencing data with computational modeling, in order to predict the function of individual bacterial strains and whole bacterial communities. Already, microbial engineering is a mature technology that exploits the power of computational modeling to optimize strains and communities for industrial processes like bioremediation and fermentation. Regarding the study of the rhizosphere microbiome, over the past 15 years this field has accumulated a huge volume of sequencing data, but these data are generally descriptive, and to date they have not been used to significantly further our mechanistic understanding of nutrient exchange processes in the rhizosphere.

One of the principal aims of mathematical models is the reduction of complexity in order to capture the fundamental principles behind the phenomena of interest. There is nowadays a positive trend in biology to develop predictive models of the system under study, with some disciplines at a more advanced stage than others in the integration of theory and experiments. The study of microbial communities is a clear illustrative example of a field where computational approaches are essential. While bioinformatics is instrumental to analyze high throughput data from meta-omics experiments, theoretical models are developed to gain mechanistic understanding of complex biological systems. Mathematical descriptions of population dynamics were pioneered in the 19th century by Verhulst's law of logistic growth and are now a fundamental part of ecology. In the same way, over the last five decades models of metabolism have started converging into sound mathematical methods to investigate cellular functions. It is also relevant to point out that another characteristic of mathematical models is that, as long as the described mechanism holds true, they can be easily generalized to different organisms. Therefore, as an example, a model originally built to describe bacteria might be also applied to other microbes.

Theoretical biology is a rapidly expanding field and it is nearly impossible to condense and classify it in few words, but we can roughly point out three main classes of methods: kinetic models, stochastic models, and network-based models. It is important to point out that depending on the system and phenomena under study, certain modeling techniques will be more suited than others. Kinetic models are dynamic and deterministic, typically constructed as systems of differential equations, solved nowadays with computational integration algorithms. These models can offer precise predictions but require either a priori knowledge or inference (e.g., through a fit to data) of the equation parameters. Typically, these models are well suited to describe small-scale metabolic pathways where most enzyme kinetic constants are measured and few parameters are reasonably constrained and then fit. Stochastic models encompass any representation that implements some random components with a Monte Carlo procedure and they are needed to capture effects like noise or individual variability. A widely used method is the Gillespie algorithm which simulates in randomized steps the evolution in time of, e.g., a chain of biochemical reactions, each of which has an associated probability to happen. Under the general term network-based models we include different static methods that have in common the treatment of metabolic pathways as networks, where metabolites are connected to reactions either as substrates or products. Reactions can be represented either as edges connecting nodes (the metabolites) or, in bipartite networks, as a disjoint set of nodes (the enzymes) that connect to the metabolites nodes via edges carrying, e.g., stoichiometric information. Reversibility of a reaction is determined by physical principles and is taken into account in directed networks.

Mostly, systems of plant–microbes have been considered in light of host–pathogen interactions. The classic zig-zag model is an illustrative scheme proposed to explain the function of the plant immune system in response to pathogens. It distinguishes two branches of the plant immune system, one reacting to microbe-associated molecular patterns with pattern-triggered immunity (PTI), the other responding to effectors trying to suppress PTI with effector-triggered immunity. This picture is, however, a purely expository model and is not suited to capture the interaction dynamics we are interested in. Indeed, if the objective is to build a predictive model of plant–microbe interactions, and in particular beneficial ones, the fundamental ingredients include: (i) the actual molecular factors driving the interactions, (ii) environmental conditions inducing different reactions, e.g., to stresses; (iii) temporal and spatial scales of the phenomena under study. While it would be nice to have a single model drawn from a universal theory of plant–microbe interactions, we are mostly at the stage where mathematical models are first needed to answer a limited number of focused and well-defined biological questions. With the current advancement in "omics" experimental techniques and with the development of computational methods to understand metabolic pathways, quantitative models of plant–microbes ecosystems at the molecular level become an appealing possibility.

The temporal scale is an intrinsic property of each biophysical process and therefore sets specific limits on the mathematical method to use. On the other hand, the spatial scale to investigate is chosen by the modeler based on the biological question to address, since we can chose to describe the same microbial community as a single metabolic unit or as a population of individual organisms. These choices are critical since they will determine at the same time the degree of complexity of the model (an Earth-wide ecosystem model can be simpler than a model of a Escherichia coli metabolism) and the requirements for the integration of experimental data.

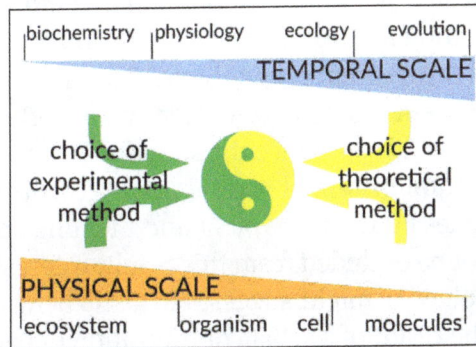

Considerations for combining modeling and experimental approaches. A major goal in biology is to integrate computational predictions with experimental data to generate predictive models of biological systems. While the temporal scale is an intrinsic property of the phenomena under study, the choice of the physical scale of the model/experiment is chosen by the scientist: one can investigate soil microbial communities at an ecosystem macroscale or at the DNA sequencing level. In general, the physical scale will have a strong impact on the choice of experimental technique, while the temporal scale will mostly influence the experimental design (this is represented by thicker/thinner arrows). On the contrary, theoretical methods will be highly constrained by the temporal scale to be modeled, while they are rather flexible regarding the physical scale to be operated at. In a truly interdisciplinary approach, the experimental and theoretical methods have to be planned together to ensure that the reciprocal results are compatible and can be integrated. This will allow further improvements in both experimental design and model development.

Dynamic Models of the Rhizosphere

The rhizosphere comprises the shared environment between the plant roots and microbes. Understanding the dynamics of nutrient competition in the rhizosphere is an important aspect to identify the driving mechanisms of community assembly. The work of Darrah presents one of the first differential equation models of microbial population dynamics in the rhizosphere, parameterizing how root growth and exudation of soluble C fuels microbial growth at specific sites. These models accounted for root elongation and release of soluble organic compounds which diffuse through the surrounding area populated by bacteria. The simulations showed how different exudation patterns can control the distribution of bacterial biomass. Being based on simple equations representing the spatial gradient of soluble substrates, this model is easily expanded to include the presence in the rhizosphere of other solutes like minerals essential to plant growth. Recently, a similar approach was used to model the colonization of the root tip by bacteria, introducing bacterial motility and under the assumption that carbon is the growth-limiting nutrient.

Borrowing concepts from microeconomics, Schott et al. modeled the nutrient trade system between plants and arbuscular mycorrhizal fungi as a minimal network, and investigated the partnership relation in terms of costs and revenues. First, they demonstrated that a simple model where the nutrient exchange via the periarbuscular space is only regulated by proton pumps (H+-ATPase) and nutrient transporters (H+/sugars and H+/phosphate cotransporters) could reproduce the situation where C and P exchange is stable at the thermodynamic equilibrium. This result shows that this small set of transporters can explain C and P fluxes between the plant and the fungi, although the presence of further transporters cannot be excluded from this result. Interestingly, while the observable highly cooperating behavior might suggest altruistic behavior in a "reward-based" system, the symbiosis is the result of a selfish optimization of personal benefit. A change in the costs to revenues ratio, for example, a well-fertilized plant that places a lower value on fungal-derived P would therefore lead to a different nutrient exchange rate.

Soil nutrient fluxes also play a role in more global scenarios. Earth system models are extended climate models that include the effects from biogeochemical cycles, and different competition theories can be embedded in them. Recently, Zhu et al. demonstrated that a kinetic model considering plant, microbes and nutrients as a coupled network, and based on the principle that nutrient uptake rate is controlled by specialized transporter enzymes, can reproduce data of N competition in a grassland ecosystem where other competition theories fail.

Genome-Scale Metabolic Network Models

Differential equation models are perfectly suited to simulate dynamic processes like enzymatic or diffusion reactions, but rely on fixed parameters that characterize specific states. In principle it would be possible to obtain exact kinetic models of metabolic networks at the "genome scale" (genome-scale models, GEMs) if we were to know the function and the kinetic parameters of each enzyme in the picture. Such models would be, however, computationally expensive to solve and difficult to manage in terms of number of equations, variables and parameters. Instead, a convenient approach is the reconstruction of GEMs from automated genome annotation and their representation as networks. In an automated process of mining the information from databases, the enzymes encoded in the genome are assigned metabolic functions. The resulting GEM, however, still requires subsequent careful manual curation as a consequence of possible inaccurate annotations besides our incomplete knowledge of genes associated to metabolic activities.

Once a metabolic network has been reconstructed, it can be translated into a stoichiometric matrix. This mathematical formalism, connecting reactions to metabolites via the stoichiometric coefficients, presents itself to a number of methods developed to study the function of biochemical pathways. Constraint-based models investigate the distribution of reaction fluxes along the directed metabolic network (first constraint, imposing irreversibility or reversibility under physiological conditions from thermodynamics principles) under the assumption that the system is at a steady state (second constraint, meaning that there is internal balance of production and consumption of metabolites) and that enzymes operate at limited capacities (third constraint, setting upper or lower bounds to the rate of an enzymatic reaction based on experimental observations). This set of constraints leads to a narrower space of possible solutions for the metabolic fluxes, but a further step is needed in order to identify a unique solution. A widely used method is flux balance analysis (FBA), where an optimization problem is defined by introducing an objective function (usually a linear combination of metabolic rates) either to be maximized (like in the case of biomass as objective function) or minimized (like in the case of total sum of cellular fluxes).

GEMs are in general compartmentalized according to the cellular structure. The simplest division, typically for prokaryotic organisms, is between an external and an internal (cytosolic) compartment, while plant GEMs will have compartments for different tissues as well. Exchange of metabolites between compartments happens through transport reactions, which are not always correctly annotated and often added to the network during the gap-filling process of the reconstruction. An exchange flux with the external compartments is equivalent to an uptake or secretion rate.

The GEM of the nitrogen fixing bacterium Rhizobium etli, a known symbiont of legumes, was studied with constraint-based modeling to understand its physiological capabilities during the third developmental stage of the symbiosis in the plant nodule, when it differentiate into a bacteroid and provides N to the plant. The authors explored how N fixation can be influenced by succinate (the carbon source provided by the plant to the bacteroid) and oxygen (nodules protect the nitrogenase enzyme from oxidative damage by providing a microaerobic environment). For a given oxygen uptake rate, an increase in succinate uptake rate leads to higher symbiotic N fixation, until a threshold value in succinate uptake rate is reached. After that an inhibitory effect, caused by insufficient oxygen for succinate reduction, drastically lowers N fixation rate. These kinds of studies, quick to perform in silico, can be easily transferred to other pathways of interest to generate hypothesis relating environment to metabolism.

Available GEMs can be browsed at the BioModels database, where most of the reconstructed models are deposited and usually classified according to their level of curation. We have to point out that efforts are still on-going to converge on model standards, and the trustworthiness of simulation results depends on the quality of the GEM.

From Single Species to Soil Communities

Even without considering the associated plant, mathematical modeling of microbial consortia is a very diverse research field where multiple methods can (and should) be applied. For a wide overview we point the interested reader to specific reviews, while here we will focus on metabolic models, pointing out advantages and limitations. A clear asset of methods like FBA is the possibility of manipulating GEMs with ease, obtaining direct information on active reactions. There are straightforward observable quantities that can be experimentally measured to inform or to challenge the model prediction, like nutrient uptake, oxygen levels, or biomass growth. However, as mentioned before, the simulation results are determined by the choice of an optimality principle, which is still an extremely subjective and approximate choice. If the concept of optimality is debatable for a single organism undergoing metabolic regulation by circadian rhythms and environmental factors, is it easy to imagine that defining optimality in a microbial community, where many other dynamics enter the picture, is far from trivial.

Considering the temporal scale of constraint-based models, the steady state condition implies that we obtain a static representation of the mass-balanced metabolic system. If we want to recover some dynamic information, we can consider that metabolism adjusts quickly to small perturbations and that it will be at a "quasi-steady-state" compared to slower external processes. Moving to the spatial scale, there is no unique way to build a community metabolic network model. Generally speaking, we can consider the results on a single organism metabolic model as a proxy of a colony of a single species, and different approaches have been proposed to describe communities of various species. For example, one possibility is to lump the metabolic pathways of each species together, obtaining a single "super-organism" where the biochemical reactions are acting together to optimize a common objective function. Another approach is to build a compartmentalized model, where each species is a separate compartment and transport fluxes allow the exchange of metabolites between them, and the modeler can choose to implement a common objective function or differentiate it among the organisms. While both choices are valid, one strategy can be more appropriate than the other in the specific context drawn by the biological question under investigation.

The recent work of Pfau is, to our knowledge, the first example of a metabolic network model of a plant–microbe system. The authors first reconstructed the GEM of Medicago truncatula, an annual legume used as model plant for studies on legume–rhizobia symbiosis, obtaining a highly curated multi-tissue model describing root and shoot. The metabolic network model of the nitrogen fixing symbiont Sinorhizobium meliloti was then connected to the root tissue model through exchange reactions extracted from literature. Simulations on the GEM of plant and symbiont under different N availability quantified the benefit, in terms of growth, for the symbiotic partners. The insight on exchange fluxes showed that S. meliloti exported only alanine as N source to the plant. A simulation repeated for an alanine dehydrogenase knock-out version of the bacterium showed that N was provided to the plant as ammonia, with an increased need for oxygen. These first results, backed-up by experimental data, already show the potential of theoretical models to provide testable hypotheses and mechanistic understanding of symbiotic interactions involving nutrient exchange.

Carbon Cycle

Carbon is the most important element in the biological system and constitutes about 50% of all living organisms. Carbon dioxide present in the atmosphere or dissolved in water is the ultimate source of organic carbon compounds occurring in nature; its complete cycle is schematically represented in figure.

The cycle of carbon in nature comprises of two main processes:

1. The conversion of oxidized form of carbon into reduced organic form by photosynthetic organisms,

2. Restoration of original oxidized form through mineralization of the organic form by the micro-organisms.

Conversion of Oxidized form of Carbon (CO_2) into Reduced Organic Form: CO_2 is reduced into organic carbon compounds mainly by the process of photosynthesis. Photosynthetic algae and higher plants are the most important agents of carbon dioxide fixation. In the ocean the major plant forms that fix carbon are the free floating microscopic algae called phytoplanktons. They are estimated to fix annually about 1.2×10^{10} tons of carbon.

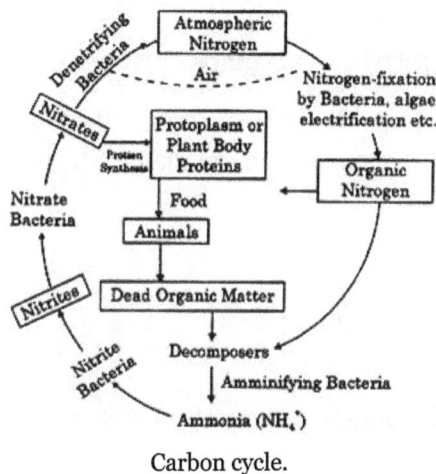

Carbon cycle.

Nearly 1.6×10^{10} tons of carbon is said to be fixed annually by photosynthetic terrestrial plant life. Besides, autotrophic and heterotrophic bacteria are also capable of synthesizing organic matter from inorganic carbon. In addition to the occurrence of photosynthesis among microorganisms, the latter also represent the example of CO_2 fixation into organic compounds which are as follows.

1. The carbon dioxide represents the sole source of carbon for autotrophic bacteria. The latter fix CO_2 to carbohydrates by a reduction reaction.

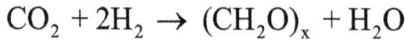

$$CO_2 + 2H_2 \rightarrow (CH_2O)_x + H_2O$$

2. Heterotrophic bacteria fix carbon dioxide commonly.

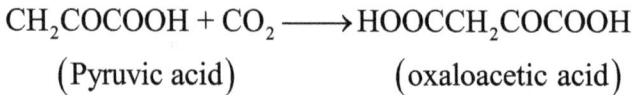

$$CH_2COCOOH + CO_2 \longrightarrow HOOCCH_2COCOOH$$
$$\text{(Pyruvic acid)} \qquad \text{(oxaloacetic acid)}$$

Restoration of Original Oxidized Form (CO_2) through Mineralization of the Organic Form:

One can consider three different modes through which the organic matter is mineralizes and the CO_2 is released in the atmosphere.

They are:

1. Process of respiration,

2. Accidental (forest fire) and intentional (fuel) burning.

3. Decomposition of organic matter by microorganisms.

The process of respiration in plants and animals, and the accidental and intentional burning of plants and their parts result in the breakdown of organic carbon compounds releasing carbon dioxide in the atmosphere.

Decomposition of Organic Matter by Microorganisms

The organic carbon compounds that eventually are deposited in the soil are degraded by the activities of microorganisms which are mainly the bacteria and fungi. The CO_2 is released into the air and soil.

(i) Cellulose Decomposition:

Cellulose is the most abundant organic material in plants. It is readily attacked by many species of fungi and bacteria.

The process of cellulose decomposition to carbon dioxide can be summarized in the form of following reactions.

1. Cellulose $\xrightarrow[\text{cellulase}]{\text{enzyme}}$ Cellobiose

2. Cellodbiose $\xrightarrow[\text{â-glucosidase}]{\text{enzyme}}$ Glucose

3. Glucose $\xrightarrow[\text{of many microbes}]{\text{enzyme system}}$ CO_2 + H_2O + and/or other end products.

The fungi which decompose cellulose in soil are mainly Trichoderma, Aspergillus, Penicillium, Fusarium, Chaetomium, Verticillium. Rhizoctonia, Myrothecium, Merulius, Pleurotus, Fomes, etc.

The bacteria that bring about cellulose decomposition in soil consist mainly of the species of Clostridium, Cellulomonas, Streptomyces, Cytophaga, Bacillus, Pseudomonas, Nocardia, Micromonospora, Sporocytophaga, Polyangium, Cellfalcicula, etc.

(ii) Hemicellulose Decomposition:

Hemicelluloses are the polymers of simple sugars such as pentoses, hexoses and uronic acid. The decomposition of hemicelluloses by microorganisms takes place through the agency of extracellular enzymes called hemicellulases.

The fungi that degrade hemicelluloses in soil are examplified by Chaetomium, Aspergillus, Penicillium, Trichoderma, Fusarium, Humicola, etc. Bacillus, Pseudomonas, Cytophaga, Vibrio, Erwinia, Streptomyces, Actinomyces, etc. are the bacteria that degrade hemicelluloses in soil.

(iii) Lignin Decomposition:

Lignin is the third most abundant constituent of the plants. It is highly resistant of microbial degradation. However, certain fungi (examplified by Aspergillus, Penicillium, Fusarium, Lenzites, Clavaria, Polyporus, etc.) and bacteria (examplified by Streptomyces, Nocardia, Flavabacterium, Xanthomonas, Pseudomonas, Micrococcus, etc.) are known to degrade lignin at slow rates.

Nutrient Cycling

Soil microorganisms exist in large numbers in the soil as long as there is a carbon source for energy. A large number of bacteria in the soil exists, but because of their small size, they have a smaller biomass. Actinomycetes are a factor of 10 times smaller in number but are larger in size so they are similar in biomass to bacteria. Fungus population numbers are smaller but they dominate the soil biomass when the soil is not disturbed. Bacteria, actinomycetes, and protozoa are hardy and can tolerate more soil disturbance than fungal populations so they dominate in tilled soils while fungal and nematode populations tend to dominate in untilled or no-till soils.

There are more microbes in a teaspoon of soil than there are people on the earth. Soils contain about 8 to 15 tons of bacteria, fungi, protozoa, nematodes, earthworms, and arthropods.

Relative Number and Biomass of Microbial Species at 0–6 Inches (0–15 cm) Depth of Soil		
Microorganisms	Number/g of soil	Biomass (g/m²)
Bacteria	$10^8 - 10^9$	40–500
Actinomycetes	$10^7 - 10^8$	40–500
Fungi	$10^5 - 10^6$	100–1500
Algae	$10^4 - 10^5$	1–50
Protozoa	$10^3 - 10^4$	Varies
Nematodes	$10^2 - 10^3$	Varies

Microbial Soil Organic Matter Decomposition

Organic matter decomposition serves two functions for the microorganisms, providing energy for growth and suppling carbon for the formation of new cells. Soil organic matter (SOM) is composed of the "living" (microorganisms), the "dead" (fresh residues), and the "very dead" (humus) fractions. The "very dead" or humus is the long-term SOM fraction that is thousands of years old and is resistant to decomposition. Soil organic matter has two components called the active (35 percent) and the passive (65 percent) SOM. Active SOM is composed of the "living" and "dead" fresh plant or animal material which is food for microbes and is composed of easily digested sugars and proteins. The passive SOM is resistant to decomposition by microbes and is higher in lignin.

Microbes need regular supplies of active SOM in the soil to survive in the soil. Long-term no-tilled soils have significantly greater levels of microbes, more active carbon, more SOM, and more stored carbon than conventional tilled soils. A majority of the microbes in the soil exist under starvation conditions and thus they tend to be in a dormant state, especially in tilled soils.

Dead plant residues and plant nutrients become food for the microbes in the soil. Soil organic matter (SOM) is basically all the organic substances (anything with carbon) in the soil, both living and dead. SOM includes plants, blue green algae, microorganisms (bacteria, fungi, protozoa, nematodes, beetles, springtails, etc.) and the fresh and decomposing organic matter from plants, animals, and microorganisms.

Soil organic matter can be broken down into its component parts. One hundred grams (g) or 100 pounds (lbs) of dead plant material yields about 60–80 g (lbs) of carbon dioxide, which is released into the atmosphere. The remaining 20–40 g (lbs) of energy and nutrients is decomposed and turned into about 3–8 g (lbs) of microorganisms (the living), 3–8 g (lbs) of non-humic compounds (the dead), and 10–30 g (lbs) of humus (the very dead matter, resistant to decomposition). The molecular structure of SOM is mainly carbon and oxygen with some hydrogen and nitrogen and small amounts of phosphorus and sulfur. Soil organic matter is a by-product of the carbon and nitrogen cycles.

Soil Organic Matter Nutrients

The nutrients in the soil have a current value of $680 for each 1 percent SOM or $68 per ton of SOM based on economic values for commercial fertilizer. SOM is composed of mostly carbon but associated with the carbon is high amounts of nitrogen and sulfur from proteins, phosphorus, and potassium. SOM should be considered like an investment in a certificate of deposit (CD). Soils that are biologically active and have higher amounts of active carbon recycle and release more nutrients for plant growth than soils that are biologically inactive and contain less active organic matter. Under no-till conditions, small amounts of nutrients are released annually (like interest on a CD) to provide nutrients slowly and efficiently to plant roots. However, with tillage, large amounts of nutrients can be released since the SOM is consumed and destroyed by the microbes. Since SOM levels are slow to build, the storage capacity for nutrients is decreased and excess nutrients released are often leached to surface waters. SOM is a storehouse for many plant nutrients.

Soils typically turnover 1 to 3 percent of their nitrogen stored in SOM. Tilled or unhealthy soils release a lower percent of nitrogen due to lower microbial activity. A tilled soil with 2 percent SOM (2,000 lbs of N) may release 1 percent N or 20 lbs of N per year. A soil that is more biologically active and has 4 percent SOM (4,000 lbs N) may release 1.5 percent N or 60 lbs N while a 6 percent SOM soil (6,000 lbs N) may release 2 percent N or 120 lbs of N. In tilled soils, excess nutrients released are often lost and the carbon stores are depleted so that future storage of nutrients is reduced. Farmers often see this occur when they till a virgin soil, an old pasture, or a fence row. For several years, crops on the newly tilled soil will grow better than the surrounding soils, but over time the soil will be depleted of carbon and the newly tilled soil will become less fertile because the carbon is oxidized as carbon dioxide and lost to the atmosphere. Tillage results in the oxidation and destruction of carbon in the soil by increasing the soil oxygen levels, thereby promoting bacteria populations to expand and consume active carbon in the soil.

Value of Soil Organic Matter	
Assumptions: 2,000,000 pounds soil in top 6 inches	
Nutrients	1% organic matter = 20,000# 50% Carbon, C:N ratio = 10:1
Nitrogen:	1000# * $0.50/#N = $500
Phosphorus:	100# * $.70/#P = $70
Potassium:	100# * $0.40/#K = $40
Sulfur:	100# * $0.50/#S = $50

Carbon:	10,000# or 5 ton * $4/Ton = $20
Value of 1% SOM Nutrients/Acre	= $680
Relative Ratio of Nutrients:	100 Carbon/10 Nitrogen/ 1 Phosphorus/1 Potassium/1 Sulfur

Climate, Temperature and pH effects on SOM

SOM is affected by climate and temperature. Microbial populations double with every 10 degree Fahrenheit change in temperature. If we compare the tropics to colder arctic regions, we find most of the carbon is tied up in trees and vegetation above ground. In the tropics, the topsoil has very little SOM because high temperatures and moisture quickly decompose SOM. Moving north or south from the equator, SOM increases in the soil. The tundra near the Arctic Circle has a large amount of SOM because of cold temperatures. Freezing temperatures change the soil so that more SOM is decomposed then in soils not subject to freezing.

Moisture, pH, soil depth, and particle size affect SOM decomposition. Hot, humid regions store less organic carbon in the soil than dry, cold regions due to increased microbial decomposition. The rate of SOM decomposition increases when the soil is exposed to cycles of drying and wetting compared to soils that are continuously wet or dry. Other factors being equal, soils that are neutral to slightly alkaline in pH decompose SOM quicker than acid soils; therefore, liming the soil enhances SOM decomposition and carbon dioxide evolution. Decomposition is also greatest near the soil surface where the highest concentration of plant residues occur. At greater depths there is less SOM decomposition, which parallels a drop in organic carbon levels due to less plant residues. Small particle sizes are more readily degraded by soil microbes than large particles because the overall surface area is larger with small particles so that the microbes can attack the residue.

A difference in soil formation also occurs traveling east to west across the United States. In the east, hardwood forests dominated and tree tap roots were high in lignin, and deciduous trees left large amounts of leaf litter on the soil surface. Hardwood tree roots do not turn over quickly so organic matter levels in the subsoil are fairly low. In forest soils, most of the SOM is distributed in the top few inches. As you move west, tall grassland prairies dominated the landscape and topsoil formed from deep fibrous grass root systems. Fifty percent of a grass root dies and is replaced every year and grass roots are high in sugars and protein (higher active organic matter) and lower in lignin. So soils that formed under tall grass prairies are high in SOM throughout the soil profile. These prime soils are highly productive because they have higher percentage of SOM (especially active carbon), hold more nutrients, contain more microbes, and have better soil structure due to larger fungal populations.

Carbon to Nitrogen Ratio

The break down of organic residues by microbes is dependent upon the carbon to nitrogen (C:N) ratio. Microbes in a cow's rumen, a compost pile, and soil microbes rely on the C:N ratio to break down organic (carbon-based) residues. Consider two separate feed sources, a young tender alfalfa plant and oat or wheat straw. A young alfalfa plant has more crude protein, amino acids, and sugars in the stalk so it is easily digested by microbes whether it is in a cow's rumen, a compost pile, or in the soil. Young alfalfa has a high nitrogen content from protein (amino acids and proteins are high in nitrogen and sulfur), so it has a lower carbon to nitrogen ratio (less carbon, more nitrogen). However, oat and wheat straw (or older mature hay) has more lignin (which is resistant to microbial decomposition), lower crude protein, and less sugars in the stalk and a higher C:N ratio. Straw is decomposed by microbes but it takes additional time and nitrogen to break down this high carbon source.

A low nitrogen content or a wide C:N ratio is associated with slow SOM decay. Immature or young plants have a higher nitrogen content, lower C:N ratios and faster SOM decay. For good composting, a C:N ratio less than 20 allows the organic materials to decompose quickly (4 to 8 weeks) while a C:N ratio greater than 20 requires additional N and slows down decomposition. So if we add a high C based material with low N content to the soil, the microbes will tie up soil nitrogen. Eventually, the soil N is released but in the short-term the N is tied up. The conversion factor for converting N to crude protein is 16.7, which relates back to why it is so important to have a C:N ratio of less than 20.

The C:N ratio of most soils is around 10:1 indicating that N is available to the plant. The C:N ratio of most plant residues tends to decrease with time as the SOM decays. This results from the gaseous loss of carbon dioxide. Therefore, the percentage of nitrogen in the residual SOM rises as decomposition progresses. The 10:1 C:N ratio of most soils reflects an equilibrium value associated with most soil microbes (Bacteria 3:1 to 10:1, Fungus 10:1 C:N ratio).

Bacteria are the first microbes to digest new organic plant and animal residues in the soil. Bacteria typically can reproduce in 30 minutes and have high N content in their cells (3 to 10 carbon atoms to 1 nitrogen atom or 10 to 30 percent nitrogen). Under the right conditions of heat, moisture, and a food source, they can reproduce very quickly. Bacteria are generally less efficient at converting organic carbon to new cells. Aerobic bacteria assimilate about 5 to 10 percent of the carbon while anaerobic bacteria only assimilate 2 to 5 percent, leaving behind many waste carbon compounds and inefficiently using energy stored in the SOM.

Fungus generally release less carbon dioxide into the atmosphere and are more efficient at converting carbon to form new cells. The fungus generally captures more energy from the SOM as they decompose it, assimilating 40 to 55 percent of the carbon. Most fungi consume organic matter higher in cellulose and lignin, which is slower and tougher to decompose. The lignin content of most plant residues may be of greater importance in predicting decomposition velocity than the C:N ratio.

Mycorrhizal fungi live in the soil on the surface of or within plant roots. The fungi have a large surface area and help in the transport of mineral nutrients and water to the plants. The fungus life cycle is more complex and longer than bacteria. Fungi are not as hardy as bacteria, requiring a more constant source of food. Fungi population levels tend to decline with conventional tillage. Fungi have a higher carbon to nitrogen ratio (10:1 carbon to nitrogen or 10 percent nitrogen) but are more efficient at converting carbon to soil organic matter. With high C:N organic residues, bacteria and fungus take nitrogen out of the soil.

Protozoa and nematodes consume other microbes. Protozoa can reproduce in six to eight hours while nematodes take from three days to three years with an average of 30 days to reproduce. After the protozoa and nematodes consume the bacteria or other microbes (which are high in nitrogen), they release nitrogen in the form of ammonium. Ammonium (NH_4^+) and nitrates (NO_3^-) are easily converted back and forth in the soil. Plants absorb ammonium and soil nitrates for food with the help of the fungi mycorrhizal network.

Microorganism populations change rapidly in the soil as SOM products are added, consumed, and recycled. The amount, the type, and availability of the organic matter will determine the microbial population and how it evolves. Each individual organism (bacteria, fungus, protozoa) has certain enzymes and complex chemical reactions that help that organism assimilate carbon. As waste products are generated and the original organic residues are decomposed, new microorganisms may take over, feeding on the waste products, the new flourishing microbial community (generally bacteria), or the more resistant SOM. The early decomposers generally attack the easily digested sugars and proteins followed by microorganisms that attack the more resistant residues.

Cover crops supply food (active carbon like glucose and proteins) to the microbes to feed on. In the soil, there are 1,000 to 2,000 times more microbes associated with roots than are living in bare or tilled soil. The microbes in turn build SOM and store soil nutrients. Building SOM requires soil nutrients like N-P-K-S to be tied up in the soil. Winter cover crops soak up excess soil nutrients and supply food to all the microbes in the soil during the winter months rather than microbes having to use up SOM reserves for nutrients. In a conventional tilled field, soil nutrients are quickly released as SOM is burned up and the microbes and soil organisms habitat are destroyed. In a no-till field, high levels of SOM are reserves of soil nutrients which are slowly released into the soils. Adding a living cover crop to a no-till field increases active organic matter (sugars and proteins) for the soil microbes. Soil microbes have two crops to feed on instead of one crop per year. Microbes thrive under no-till conditions and winter cover crops. Cover crops and manure can be used to feed soil microbes and recycle soil nutrients. As soil microbes decompose organic residues, they slowly release nutrients back into the soil for the winter cover crops or for the preceding crop. Cover crops prevent the nutrients from being lost through soil erosion, leaching, volatilization, or denitrification.

Phosphorus Cycle

Phosphorus is one of the most important constituent of several important compounds always present in organisms. It occurs both in organic (nucleic acids, nucleoproteins, phospholipids, etc.) and inorganic (phosphate) forms in the living organisms. Animals possessing bones have large amount of phosphorus in its inorganic form.

However, phosphorus is added to soil through chemical fertilizers, excrete and organism- residues. Though there is plenty of phosphorus present in the soil in unavailable inorganic forms, most of the plants obtain it only as orthrophosphate ions (soluble inorganic forms). However, mycorrhizae, when present, help the plants in obtaining phosphorus.

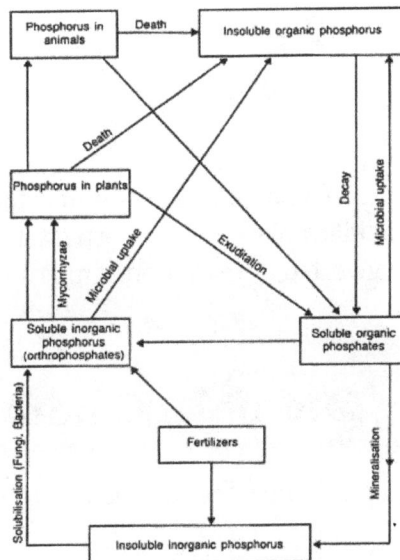

Schematic representation of phosphorus cycle.

The cycle of phosphorus is schematically represented in above figure and can be well studied under following two heads:

- Mineralization: Conversion of Organic Phosphorus into Insoluble Inorganic Phosphates:

Many soil microorganisms produce enzymes that attack many of the organic phosphorus compounds in the soil and release inorganic phosphate. This process is comparable to the mineralization of organic nitrogen compounds. The enzymes involved in these reactions are collectively called 'phosphatases' which have a broad range of substrate specificity.

- Solubilization: Conversion of Insoluble Inorganic Phosphates into Soluble Inorganic Phosphates.

The availability of phosphorus depends on the degree of solubilization by various organic and inorganic acids produced by microorganisms in soil. These are the solubilized form of insoluble inorganic phosphates which are taken in by the plants.

Fungi, e.g., Aspergillus, Penicillium, Fusarium are the most important of the soil microorganisms which produce substantial amounts of these acids; others are the bacteria namely Bacillus, Pseudomonas, Micrococcus, Flavobacterium, etc.

The overall conversion of insoluble inorganic phosphates into soluble inorganic phosphates by the action of acids can be exemplified via reactions as under:

$$Ca(PO_4)_2 + H_2SO_4 \longrightarrow Ca_2H_2(PO_4)_2 + CaSO_4$$
(Insoluble calcium phosphate) (Sulfuric acid) (Calcium-monohydrogen phosphate fairly soluble in water)

$$Ca_2H_2(PO_4)_2 + H_2SO_4 \longrightarrow CaH_4(PO_4)_2 + CaSO_4$$
(Calcium-dihydrogen phosphate: highly soluble in water)

$$Ca_3(PO_4)_2 + 2HNO_3 \longrightarrow Ca_2H_2(PO_4)_2 + Ca(NO_3)_2$$
(Insoluble calcium phosphate) (Nitric acid) (Calcium-monohydrogen phosphate: fairly soluble in water)

$$Ca_2H_2(PO_4)_2 + 2HNO_3 \longrightarrow CaH_4(PO_4)_2 + Ca(NO_3)_2$$
(Calcium-dihydrogen phosphate: highly soluble in water)

The action of acids to convert insoluble phosphates into soluble ones is generally called 'solubilization' and particularly takes place in close proximity of the root surfaces where sugar from root-exudates are converted by the action of microorganisms into organic acids.

Factors affecting Microbial Community in Soil

The major external factors that influence the microbial community in soil are:

1. Soil Moisture.
2. Organic and Inorganic Chemicals.
3. Soil Organic Matter.
4. Types of Vegetation and its Growth Stages.
5. Different Seasons.

Soil Moisture

Moisture is present in the form of film in soil pores. The amount of water increases with increase in porosity of soil. Pore-size depends on soil texture i.e. composition of sand, silt and clays. Moreover, soil moisture is affected through irrigation, drainage or management practices like tillage or crop rotation that enhance the intake and transmission of water by soil.

Organic and Inorganic Chemicals

The chemicals are very important for microorganisms as these provide nutrition for growth, activity and survival of microorganisms in ecologically deficient niches in soil. The chemical factors are gases, acids, micro- and macro-elements, clay minerals, etc. In the soil solution gases (oxygen, methane and carbon dioxide), and microorganisms are dissolved.

However, the dissolved components are in constantly shifting equilibrium with the solid phase, soil air, and moisture as well as with soil organisms and plant root activity. It has been found that low potassium and high nitrogen favour cotton wilt by Fusarium vasinfectum.

Soil-borne fungi are sensitive to pH. As a result of pH range for vigour and growth, they are more destructive at acid and neutral at alkaline conditions.

For example, Plasmodiophora brassicae favours best in acid soil, and the disease produced by it is uncommon or mild in soil of pH more than 7.5. Acidophilic natives of Trichoderma viride increased in soil on addition of sulphur, carbon disulphide, and methylbromide due to lowering down of pH to about 4.0.

Soil Organic Matter

The dead organic material of plant and animal origin serve as total soil organic matter which later is subjected to microbial colonization and decomposition. However, upon incorporation of green manures, crop residues, etc. in soil, the community size of microorganisms gets increased.

At the same time application of these organic matter alters the composition of soil microflora, microfauna, and relative dominance of antagonistic bacteria, actinomycetes, fungi, amoebae, etc.

Due to excess soil moisture in peat bogs, the growing plants cannot absorb enough minerals (as their roots become waterlogged) and therefore the microbial population is lower.

Types of Vegetation and its Growth Stages

The dominance of one or the other groups is related to the type of vegetation and growth stages of a plant. Dubey and Dwivedi found an increased population of fungi in the non-rhizosphere and rhizosphere of soybean according to season and growth stages, respectively.

In the rhizosphere aspergilli, fusaria and penicillia were dominant in addition to the other fungal species. However, frequency of Macrophomina phaseolina and Neocosmospora vasinfecta increased on rhizoplane with onset of senescence.

This selective action of plants is attributed to microbial response either to specific root-exudates or chemical constituents of sloughed-off tissues that undergo decomposition. Moreover, Mueller determined the incidence of fungi and bacteria occurring in the roots of six soybean cultivars growing in fields cropped for 3 years either with corn or soybean.

Cropping history affected the recovery of M. phaseolina, Phomopsis spp. and Trichoderma spp. but not Fusarium spp. or Gliocladium roseum. Recovery of Trichoderma spp. was greater following com than following soybean. After death of the plant soil saprophytes colonize rapidly, thus total spectrum of microflora in the rhizosphere is changed.

Different Seasons

The amount of plant available nutrients is governed by the number and activity of microorganisms. They remain in constant dynamic state in soil where microbial community is greatly influenced by physicochemical and biological factors. Changes in microbial community are known in soils of tropical, sub-tropical and temperate regions.

Shail and Dubey have studied the seasonal changes in microbial community (bacteria and fungi) and species diversity in fungi in banj-oak and chir-pine forest soils of Kumaon Himalaya in relation to edaphic factors. Maximum number of fungal taxa and average number of bacteria and fungi (per gram soil) were recorded in rainy season and minimum in summer season from both the soils.

Seasonal changes in average number of fungi and bacteria per gram dry soil in banj-oak and chir-pine forest soils of himalaya.

Forest soils	Seasons	Average number (g^{-1} dry soil)	
		Fungi (CFU $\times 10^3$)	Bacteria (CFU $\times 10^5$)
Banj-oak	Summer	199	987
	Rainy	369	1408
	Winter	189	1161

Chir-pinc	Summer	70	908
	Rainy	272	1344
	Winter	129	1072

Soil Microbial Interactions and Organic Farming

Organic farming utilizes diverse communities of microorganisms to protect plants from disease and to keep soil healthy and productive. Microorganisms such as bacteria, single-celled fungi, microalgae, and protozoa are all part of the soil ecosystem, along with the plants and animals that rely on the soil to survive. These microfora and microfauna form the foundation of an ecological web of organisms that live in the soil and play an important role in cycling nutrients that sustain plant and animal life. As agents for biological, chemical and physical change, these microorganisms transform soil properties.

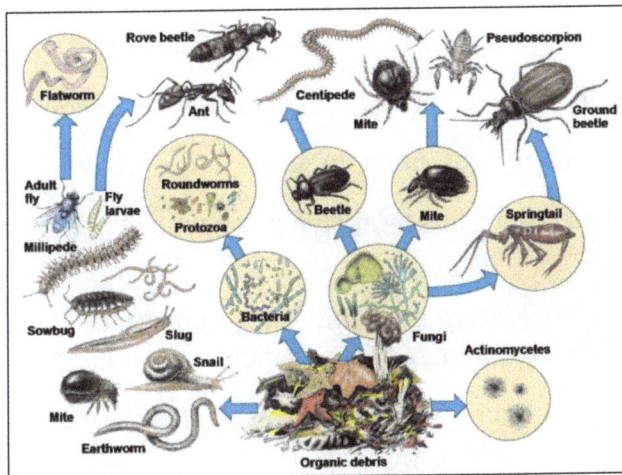

The soil food web

Soil quality can be measured in different ways. Healthy soil has diverse populations of microorganisms and microfauna, suffcient organic matter, and is resilient to disturbances like tillage, drought, and fooding. Monoculture agriculture, in combination with intensifcation and heavy chemical use, reduces the quality and biodiversity of the soil ecosystem, and thus its resilience, sustainability and long-run productivity. These farming methods can make the soil more susceptible to adverse weather conditions and infestations by pests and diseases that reduce productivity.

Better soil quality provides higher yields and greater profts for farmers over the long run. To enhance soil quality, farmers can use soil building techniques that maintain or increase the amount of organic matter in the soil. These techniques include growing cover crops and adding compost. Crop rotation and intercropping also play an important role in diversifying the soil ecological system over time.

Most soil quality indicators looked at the chemical and physical characteristics of the soil. Today, we have a better understanding of soil ecology and know that soil health can also be measured by the life within it. While chemical and physical characteristics of the soil change slowly from year to year, soil biology has dynamic properties with implications for soil chemistry and structure. Populations of selected species, microbial biomass, respiration rates, and enzymatic activity can be used to measure total soil microbiological activity and estimate the ecosystem services provided by soil microorganisms.

There are several methods for regenerating soil health, including soil building for plant health and nutrition, biological remediation, and the introduction of benefcial soil organisms. These methods all complement each other and use the same principles of soil ecology.

Best Practices for Soil Health

- Maintain and build soil organic matter by growing cover crops and applying compost.

- Rotate annual crops to alternate hosts of pathogens.

- Where possible, intercrop or grow mixed crops to provide a diversified habitat.

- Plant disease-free, uncontaminated seed.

- Replace harvested nutrients by applying amendments from plants, animals and minerals.

- Avoid highly soluble sources of nutrients, particularly those that have a high salt index or are contaminated with elemental contaminants.

- When growing legumes, inoculate with rhizobial bacteria symbiotic with the host plant.

- Inoculate seeds with treatments comprised of benefcial microorganisms that protect seedlings from soil borne diseases.

Plant Health and Nutrition

Plants require air, water and nutrients to survive, but the biological basis for plant health goes beyond survival and productivity. Most terrestrial plants have adapted to grow in soil conditions with symbiotic, mutualistic, and parasitic microorganisms. Although plant pathology has focused on the parasitic or pathological interactions, the benefcial relationships of plants with microorganisms have long been recognized. One of the most studied interactions is that of nitrogen-fxing rhizobial bacteria that have a synergistic relationship with legumes. These bacteria form nodules on the roots of plants in the bean and pea family, which in turn capture nitrogen from the atmosphere and make it available to the plants.

A healthy, balanced soil ecosystem provides a habitat for crops to grow without the need for interventions such as soil fumigants and drench pesticides. Replacing and building soil organic matter enables sustainable cycling of nutrients, like nitrogen. Sources of soil organic matter include cover crops, compost, and other soil amendments that come from plants and animals. Cover crops are grown and turned back into the soil without being harvested. Compost is decomposed organic matter teeming with microorganisms that cycle nutrients to build healthy soil. Soils rich in organic matter offer the benefit of reduced leaching and volatilization of soil nitrogen.

Soil microorganisms also make other nutrients more bioavailable through the decomposition of organic matter and biochemical reaction with the mineral portion of the soil.

Vesicular-arbuscular mycorrhizae (VAM) are microorganisms that have specific symbiotic functions with plants. These microscopic fungi attach themselves to the roots of plants and exude various enzymes and other chemicals that make soil nutrients more plant available.

Nutrients become biologically available (bioavailable) when released from the mineral or organic fraction of the soil to be taken up by plant roots and other organisms in the soil. In the case of crop nutrition, bioavailability is a good thing, helping the soil to feed the plants. In the case of soil contamination with heavy metals, bioavailability may be harmful, in that the contaminants can be taken up by the plants and the animals and humans that consume those plants.

Bioremediation

Biological remediation is necessary in agricultural soils because conventional farming practices using chemical fertilizers and pesticides, industrial pollution, and atmospheric deposition pollute the soil. Some contaminants can persist and have adverse effects on plants and animals for years. As part of the process of transitioning to organic agriculture, farmers face the challenge of remediating polluted soil.

Bioremediation is an intentional process used to eliminate environmental pollutants from contaminated sites. Other contaminants that can by removed via bioremediation include those from industrial processes and hazardous waste. The process includes using microorganisms, fungi and plants to decompose, sequester or remove soil contaminants, such as agricultural pesticides and fertilizers.

Seedling

Through a series of complex metabolic interactions, often involving a number of different organisms, unwanted contaminants can be broken down, immobilized or removed. Organic farmers can choose the appropriate bioremediation technique for their farm based on the pollutants to be removed, soil types, weather conditions, and biome.

One technique involves introducing organisms that biodegrade, make bioavailable, or sequester different contaminants. There are also a number of microbial inoculants for introducing benefcial bacteria or fungi to seeds and soil. Plants grown as cover or catch crops can also help with the breakdown or removal of various soil-borne contaminants.

Soil amendments can change the mix of organisms as well. For example, amendments with hard, slow to break down surfaces—like crab shells—have a layer of what is known as 'chitin' that needs to decompose. The microorganisms that decompose chitin can also suppress plant-eating nematodes by dissolving their shells. Sesame chaff also suppresses pathogenic nematodes.

However, there are situations where bioremediation can make a contamination situation worse by making a sequestered contaminant more bioavailable. Therefore, laboratory experimentation is helpful before trying an approach in the feld.

For example, elemental contaminants—such as arsenic, lead, and cadmium—are not biodegradable and can only be sequestered and removed. Treatments combine microorganisms to make the contaminants bioavailable for plants that are heavy accumulators of those contaminants. Once the plants absorb the contaminants, the plant biomass needs to be removed, otherwise the contaminants will be concentrated in the organic matter in a relatively bioavailable form.

While most carbon-based pollutants will break down more quickly with greater biological activity; some pollutants, such as chlorinated hydrocarbons, actually become more persistent with higher levels of organic matter. With so many different contaminants, environmental conditions, and possible approaches, it may take some trial and error to fnd a successful approach to remediating a given contamination situation. Bioremediation is becoming more cost effective compared with chemical and physical approaches, which are increasing in costs and also have their drawbacks.

Microbial Benefits for Disease and Weed Control

Under ordinary conditions, soil-borne pathogens play an important role in agricultural ecosystems by helping with the decay of plant tissue. Severe outbreaks of disease are themselves symptoms of imbalance in a system, whether it is a nutrient excess or defciency, lack of genetic diversity, or a monoculture system that provides too great a host population. Biological control using soil borne organisms works by several different modes of action. These include competitive exclusion, hyperparasitism, the production of natural antibiotics, systemic acquired resistance, and induced systemic resistance.

In competitive exclusion, one organism creates an environment that is unwelcoming for another organism, effectively excluding the second organism from becoming established without directly killing it. An example of this would be the creation of flm on a root surface to prevent pathogens from infecting the plant.

The same mycorrhizae that help make nutrients biologically available can also be antagonistic for a number of opportunistic plant pathogens. Some of these microorganisms produce natural antibiotics, like Streptomyces, which became a commercial source for antibiotics. These antibiotics work to suppress different pathogens in the soil the same way they do in people and animals; the microorganisms produce the antibiotic that kills the exposed pathogens.

Specifc organisms are known to protect seeds and seedlings from various diseases. For instance, various Bacillus, Trichoderma and Pseudomonas species protect the roots of plants from infectious diseases. These organisms can be introduced by inoculation of the soil. The effcacy of soil inoculants varies widely. Biological control inoculants are not as well understood as nitrogen fxing rhizobial inoculants, and the circumstances where they work and do not work is a subject of ongoing investigation.

The Organic Farming Research Foundation (OFRF) has funded three projects researching the ability of benefcial microbes to aid organic farming. Carr investigated the benefts of using compost to introduce microbes that would protect seedlings against damping off disease caused by a Pythium species. Nebert looked at seed disinfection and inoculation as a way to avoid or prevent seed-borne diseases, particularly Fusarium. Zinati examined the use of compost extract as a means to inhibit weed seed germination and reduce weed competition. These projects highlight the potential for new research to explore the benefcial role microbes can play in organic plant production.

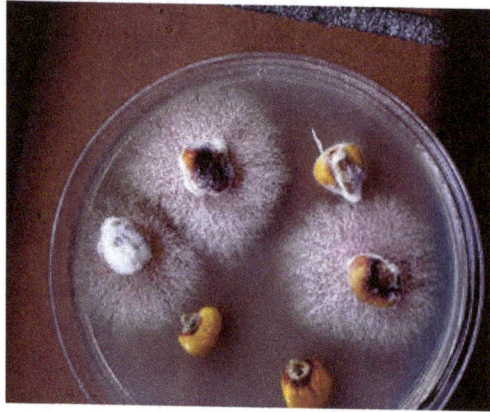

Fusarium infection on corn seeds.

Types of Soil Microbes to Nurture Plant and Soil Health

It's hard to pick just a few soil microbes from an ingredient list that contains over 800 species of diverse soil microorganisms. However, while the microbe count in Bio 800+ is important, diversity is even more crucial. Because Bio 800+ is filled with a diversity of life, by using the product you prepare the soil and plant for whatever hurdle it needs to face.

Think of it as your equivalent to a multi-vitamin. Instead of a magic bullet, Holganix Bio 800+ products parachute in an army of specialists, each of which is capable of filling in gaps to balance the biological requirements of the soil and plants.

Yeast

Species of yeast representing two genus are present Debarayomyces and Saccharomyces. Species of both these genera are known to be plant growth promoting organisms. The plant growth promoting effect of yeast in these genera include phosphate solubilizing, barriers to pathogenic bacteria and fungi, production of plant growth hormones and other effects.

Trichoderma

Trichoderma provides proteins to the plant that produce an induced systemic resistance effect in the plant (ISR). This ISR effect triggers the plant to mount disease resistance properties and also to stimulate photosynthesis in the leaves. In addition, Trichoderma species produce and release plant growth hormones to the soil.

Penicillium

Penicillium fungi have a plant growth effect caused by the production and secretion of gibberellins which stimulate root growth. Additionally, Pennicillium species have an induced systematic resistance (ISR) effect on plants by secreting enzymes that trigger a plant immune response.

Nematodes

It is often joked among scientists that nematodes can be found everywhere. About 90% of the ocean floor, and 80% of Earth carries populations of nematodes. They are even rumored to be one of the first organisms to develop from a single-celled organism to a multi-celled organism.

Within the nematode family, there are both beneficial and pathogenic microorganisms. On the one hand, beneficial nematodes are important in mineralizing nutrients (converting nutrients into plant-useable forms), but on the other hand, pathogenic nematodes can destroy root systems, cause disease and negatively impact mycorrhizal fungi. Other species of nematodes are "grazers" in the environment they live in. These grazers eat bacteria, fungi and/or other nematodes (dependent on the nematode species doing the grazing). This battle is playing out in the soil under our feet every day. Holganix Bio 800+ contains nematodes just waiting to join this battle as reinforcement upon application to the soil.

Lactobacillus

Lactobacillus is a genus of bacteria commonly found in soil and the guts of animals. In the soil, these organisms promote the decomposition of organic matter. In addition to organic matter decay, they suppress disease and perform growth-regulatory effects on fungi, yeast and other bacteria.

There are over 200 species of Lactobacillus known to science, with many more waiting to be identified as knowledge of this genus expands.

Soil Microbes in Plant Sulphur Nutrition

Chemical and spectroscopic studies have shown that in agricultural soils most of the soil sulphur (>95%) is present as sulphate esters or as carbon-bonded sulphur (sulphonates or amino acid sulphur), rather than inorganic sulphate. Plant sulphur nutrition depends primarily on the uptake of inorganic sulphate. However, recent research has demonstrated that the sulphate ester and sulphonate-pools of soil sulphur are also plant-bioavailable, probably due to interconversion of carbon-bonded sulphur and sulphate ester-sulphur to inorganic sulphate by soil microbes. In addition to this mineralization of bound forms of sulphur, soil microbes are also responsible for the rapid immobilization of sulphate, first to sulphate esters and subsequently to carbon-bound sulphur. The rate of sulphur cycling depends on the microbial community present, and on its metabolic activity, though it is not yet known if specific microbial species or genera control this process. The genes involved in the mobilization of sulphonate- and sulphate ester-sulphur by one common rhizosphere bacterium, Pseudomonas putida, have been investigated. Mutants of this species that are unable to transform sulphate esters show reduced survival in the soil, indicating that sulphate esters are important for bacterial S-nutrition in this environment. P. putida S-313 mutants that cannot metabolize sulphonate-sulphur do not promote the growth of tomato plants as the wild-type strain does, suggesting that the ability to mobilize bound sulphur for plant nutrition is an important role of this species.

Because of the increasing use of low sulphur fuels, and of enhanced emission controls, there has been a dramatic reduction in the atmospheric deposition of sulphur in recent years. This change has had an important impact in agriculture, since crop plants have become increasingly dependent on the soil to supply the sulphur that they need for the synthesis of proteins and a number of essential vitamins and cofactors. From the plant's perspective, the most important form of sulphur is inorganic sulphate, since this is the starting point for cysteine biosynthesis. However, inorganic sulphate forms only a very small part of the sulphur that is present in soils and, as a result, symptoms of sulphur deficiency are now frequently encountered in crop plants.

However, although inorganic sulphate generally makes up less than 5% of the sulphur present in agricultural soils, this does not mean that these soils contain limiting amounts of total sulphur. Most of the sulphur in soil environments (>95% of total sulphur) is bound to organic molecules, and is therefore not directly plant-available. This organic sulphur is present as a heterogeneous mixture of forms, partly included in microbial biomass and partly in the soil organic matter, and very little is known about the chemical identity of the specific sulphur-containing molecules. Traditionally, the types of sulphur species have been differentiated by their reactivity to reducing agents, allowing the organosulphur pool to be divided up into three groups: (i) HI-reducible sulphur (thought to be primarily sulphate esters); (ii) Raney-nickel-reducible sulphur; and (iii) residual carbon-bonded sulphur (thought to be largely sulphonates and heterocyclic sulphur). The identity of these groups has recently been confirmed by an independent method, using X-ray near-edge spectroscopy. Using XANES, the oxidation state and co-ordination environment of bound sulphur in soils can be compared with standard molecules, and spectral modelling is used to estimate the proportion of each sulphur form in the tested soil. Although the results obtained with XANES and 'wet' techniques are broadly similar, there are significant differences, especially in the sulphate ester fraction. The method has great potential as a non-invasive technique, allowing detailed analysis of sulphur dynamics, but the need for a synchrotron is possibly delaying its establishment as a routine technique.

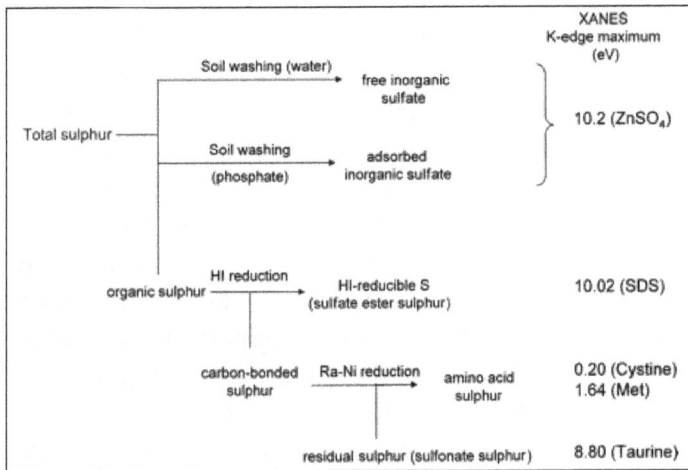

Procedures for analysis of total soil sulphur - 'Carbon-bonded sulphur' is calculated as the difference between total sulphur and HI-reducible sulphur. Similarly, residual sulphur is the calculated net organic sulphur after subtracting amino acid S and HI-reducible S. XANES K-edge values are given in eV relative to the value for elemental sulphur.

Importantly, although some of the organosulphur present in soils is plant- and animal-derived, much is also synthesized in situ. The sulphur pools in soil are not static, but extremely dynamic. Inorganic sulphur forms are immobilized to organic sulphur, different organosulphur forms are interconverted, and immobilized sulphur is simultaneously mineralized to yield plant-available inorganic sulphur. These processes occur concurrently, and many of them are linked to the microbial biomass present in the soils. Especially in the rhizosphere, it is clear that microbes play a critical role as a link in allowing plants to access soil organosulphur.

Sulphur Immobilization Processes

The two critical processes in sulphur cycling in soils, immobilization of inorganic sulphur and mobilization of organically bound sulphur, are both thought to be microbially mediated, but it is not yet known whether specific members of the microbial community play a dominant role in catalysing these processes. Immobilization has been studied in most detail using radiolabelled ^{35}S-sulphate, and measuring its incorporation into the different pools of bound sulphur. In an early study, soils incorporated 35–44% of labelled sulphate into organosulphur within a period of 8 weeks, but this and other early reports applied the radiolabel together with carrier sulphate, modelling the changes in sulphur dynamics expected on the addition of sulphate fertilizers. However, the presence of carrier sulphate was found to retard the incorporation of ^{35}S into the organosulphur pool, and later work has used carrier-free ^{35}S-sulphate. Labelled sulphate is most rapidly incorporated into the sulphate ester pool (HI-reducible sulphur), and more slowly into the carbon-bonded S fraction. Importantly, the rate at which added sulphate is immobilized depends critically on soil conditions, both on how the soil is preincubated before the addition of the ^{35}S-sulphate and on the carbon and nitrogen supplied to the soil at the time of label incorporation, implicating the soil microbial community as the major player in this process. Preincubation under moist conditions to encourage microbial growth prior to the addition of labelled sulphate leads to slower inorporation of ^{35}S-sulphate into the sulphate ester pool than when the ^{35}S-sulphate is added directly to air-dried soils, presumably because a burst of microbial growth on moistening encourages rapid sulphate immobilization. The addition of glucose as a readily utilized carbon source also encourages rapid bacterial growth, and leads to high levels of incorporation into the C-bonded S fraction (>70% of added ^{35}S in 20 d), and a similar effect is seen with organic acids (succinate/malate), although in this case only total incorporation into organosulphur was measured. Similarly, the incorporation of sulphur into the organic S pool was increased dramatically by the incorporation of cellulose as an additional carbon source, with 40–50% incorporation observed in the presence of cellulose, and only 10–20% in its absence. Microbial dry matter contains about 40% carbon and 1% sulphur, so the provision of an excess of bioavailable carbon stimulates microbial growth, but simultaneously leads to an enhanced requirement for sulphur. The addition of extra nitrogen (as ammonium nitrate) together with glucose slightly stimulated S immobilization above that of glucose alone. The stimulation of bacterial growth in these supplemented systems may be regarded as similar to the stimulation of bacterial growth that occurs in the natural rhizosphere relative to the bulk soil (the so-called 'rhizosphere effect'), due to the release of organic acids and sugars in plant root exudates. Interestingly, it is evident that soil microbes actively compete with plants for the available sulphate under these carbon-enriched conditions. In soils supplemented with cellulose, for example, plant yield and plant sulphur content were both reduced, suggesting that, over the time period studied, soil microbes were able to bind all the available sulphate into microbial biomass and thus deprive the plants of sulphur.

Mineralization of the Soil Organosulphur Pool

Although many studies of sulphate immobilization have aimed at understanding the mobility and fate of sulphate in soils, a secondary purpose has often been to generate a labelled organic S pool, in order to evaluate the rate of mineralization, or remobilization, of the bound sulphur. It is clear from several studies that the most rapidly mineralized pool of organic S is the sulphur that has been most recently immobilized, and that immobilization and mineralization are taking place concurrently. The reason for this is somewhat debated. Castellano and Dick have suggested that immobilized S makes its way initially into the sulphate ester pool, and is then slowly converted by microbial action into C-bonded S. There are also reports of both bacteria and fungi catalysing immobilization of sulphate to choline sulphate, although this does not appear to have been studied in detail. Since sulphate ester-S is intrinsically more susceptible to hydrolysis (chemical or enzymatic) than is C-bonded S, it might therefore be expected to be more readily mineralized to sulphate. Eriksen, by contrast, has classified soil sulphur into forms that differ in the ease with which sulphur may be extracted from the soil particles, and has shown that recently-formed organic S is less physically protected than aged organic S, and the latter is therefore released more slowly. Both these reports represent field studies, whereas in a laboratory study it was reported that for recently immobilized S both sulphate ester and C-bonded S were rapidly mineralized.

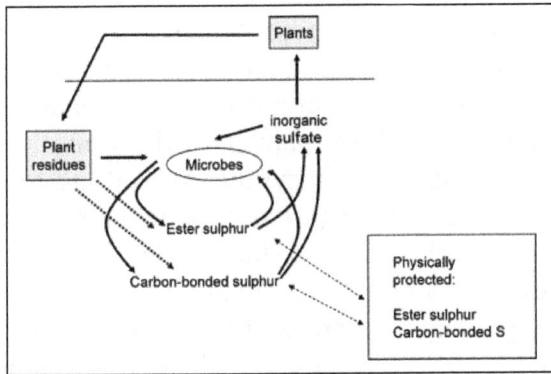

Sulphur cycling within plant–soil–microbe systems, derived from the model of McGill and Cole as elaborated by Eriksen et al. The model has been described as oversimplified, mainly because the microbial biomass acts both as a catalyst of the transformations indicated and as part of the organic sulphur pool.

Probably the clearest finding regarding organosulphur transformations in soils is that the proportions of sulphate ester sulphur and C-bonded sulphur in a given soil, and the rates in which they are interconverted and mineralized, depend critically on the cropping of the soil concerned. The role of the plant in controlling sulphur transformations in the soil is thought to derive primarily from the increased microbial biomass present in the rhizosphere compared with the bulk soil. Clear differences in sulphur transformations are observed for different plants, but until recently there has been little attempt to measure these within the rhizosphere separately from the bulk soil. Most studies have selected plant systems for investigation based on their perceived sulphur requirement, for example, a recent paper examined oilseed rape, radish, and wheat because of differences in their total S requirement and total S uptake. However, the microbial community composition of the rhizosphere varies dramatically between different plant species This effect is thought to be largely due to differences in the amount and composition of root exudates. Differences in sulphur transformations observed in soils grown with different plants may, therefore, be due to differences in the microbial community in the rhizosphere rather than purely a response to plant sulphur demand. It seems very likely that particular microbial species or genera in the rhizosphere play a greater role in sulphur cycling than others, but this functional specialization has not yet been investigated in any detail. A major stumbling block is that because most soil microbes cannot be grown in the laboratory with current techniques, cultivation-based techniques afford a very biased view of the microbial community. For a fuller under-standing of the microbial role in sulphur cycling, a targeted functional diversity analy-sis will be required, using cultivation-independent techniques.

Sulphate Esters and Microbial Soil Competence

Is soil organosulphur really necessary for soil microbes, or can they survive on the re-sidual sulphate present in the soil? Initial studies have been carried out with a strain of Pseudomonas putida, which is a typical representative of the plant-associated micro-flora. The fluorescent pseudomonads have long been regarded as rhizosphere inhab-itants par excellence, and they play an important role in many agricultural systems, since many Pseudomonas strains have plant growth-promoting effects, stimulating plant growth both directly (e.g. by auxin synthesis), and indirectly (e.g. by pathogen suppression). We chose to study rhizosphere sulphur metabolism in a Pseudomonas putida isolate, strain S-313R. This strain is characterized by its ability to utilize an ex-ceptionally broad range of organosulphur compounds as the sole sulphur source during in vitro growth, including many aromatic and aliphatic sulphonates and sulphate es-ters. A transposon mutant of this strain was isolated that was unable to desulphur-ize either arylsulphate esters or alkylsulphate esters in vitro, but grew normally with sulphonates and with sulphate (strain PH18). Its inability to grow with sulphate esters as a sulphur source was found to derive from a mutation in the SftR transcriptional activator protein, a member of the LysR-family that controls the expression of several sulphatases and sulphate ester transport systems, both in this strain and in other pseu-domonads. To assess the importance of sulphate esters for bacterial growth in soil and rhizosphere environments, survival of the wild-type strain was compared with that of the sftR mutant in microcosm experiments using three soils of different land-use his-tories from the same area (agricultural soil, forest soil, and a natural grassland soil). After incubation in washed, uncultivated soils, survival of the sftR mutant strain was significantly reduced relative to the wild-type strain. In the Arabidopsis rhizosphere the effect was even more pronounced, since the wild-type strain was able to establish itself fairly stably, while the mutant strain died off significantly over 30 d. The ability to use sulphate esters is therefore critical for bacterial survival in the soil and rhizosphere. This was also reflected in an experiment, in which 100 bacterial strains were isolated at random from garden soil on non-selective medium, subjected to sulphate starvation and tested for arylsulphatase activity using X-sulphate as substrate. Sulphatase activity was observed for all colonies tested. This contrasts with similar experiments carried out with oral bacteria, where a much smaller proportion of the isolates were sulpha-tase-positive, confirming the importance of sulphatases in the soil environment.

Arylsulphatase has been extensively studied as an important soil enzyme catalysing the hydrolysis of sulphate esters in the soil. The original model of McGill and Cole, which divided sulphur metabolism pathways in the soils into 'biological' pathways catalysed by micro-organisms and 'biochemical' pathways depending on free soil enzymes relied heavily on the idea of arylsulphatase as an enzyme secreted by bacteria into the external environment as a response to sulphur limitation. Extracellular and intracellular sulphatase activities are distinguished by measuring enzyme activity before and after treatment with a plasmolytic agent. It is interesting to note, however, that although many arylsulphatases from enteric bacteria are indeed extracellular, sulphatases identified in Pseudomonas species are almost exclusively intracellular, often coupled with active sulphate ester uptake systems. Nonetheless, total arylsulphatase activity in soils is correlated with microbial biomass, and also with the rate of S immobilization. Interestingly, the addition of exogenous arylsulphatase to soils did not appear to stimulate sulphate release from soil sulphate esters.

Plants Assimilate Carbon-bound Soil Sulphur

Sulphate is immobilized to sulphate ester-sulphate, and subsequently converted by microbial action to carbon-bonded sulphur. The microbial survival data corroborate this by suggesting that the most important organosulphur pool for soil bacteria is the sulphate ester pool. Microbial C-bound sulphur may enter the soil pool of C-bound sulphur after the death of the microorganism, or through protozoal predation releasing bacterial cellular contents into the soil environment. Although plants cannot access this C-bonded S directly, the bound sulphur in this pool is available to plants indirectly via a process that is thought to be mediated by microbial action. To test the importance of microbes in this process, a series of mutants of P. putida S-313 were generated that are unable to utilize the sulphur of aryl- and alkylsulphonates in vitro. These mutants were tested for their ability to stimulate the growth of tomato plants, after inoculation into the soil. In the presence of P. putida S-313, tomato plant growth over 26 d (measured as shoot dry weight) was increased by 1.9-fold over that of the uninoculated control plants, indicating a significant plant-growth-promoting effect by this strain. Such plant-growth-promoting effects are not uncommon: the bacteria may stimulate plant growth by increased nutrient mobilization, protection against pathogens, or production of phytohormones.

The sulphatase-deficient mutant described above (strain PH18) showed the same plant-growth-promoting effect, demonstrating that although sulphate ester utilization is important for bacterial soil competence, it is not the limiting factor in the microbial stimulation of plant growth that is observed here. (Population analysis showed that the sulphatase-negative strain was still present in the rhizosphere at the end of the test period, despite showing reduced soil competence compared with the wild type). Mutant strains that were unable to desulphurize aromatic or aliphatic sulphonates in vitro showed a different effect on plant growth: strain SN36, deficient in the asfA gene. Growth of tomato plants with either of these two strains in the rhizosphere did not lead to the growth stimulation that was observed with the wild-type strain. For the asfA mutant, plant growth was reduced to the level of the uninoculated control, whereas the ssuE mutant led to an intermediate level of plant growth stimulation. The only known role of the ssu and asf loci is in sulphonate desulphurization and their expression is regulated by sulphur supply, so it seems unlikely that they are critical in, for example, phytohormone synthesis. This suggests that the plant-growth promoting effect of this strain is directly related to its ability to mobilize sulphonate-sulphur, but it remains to be clarified whether P. putida S-313 can deliver sulphonates directly to the plant, or whether the sulphonate-sulphur first enters the sulphate pool.

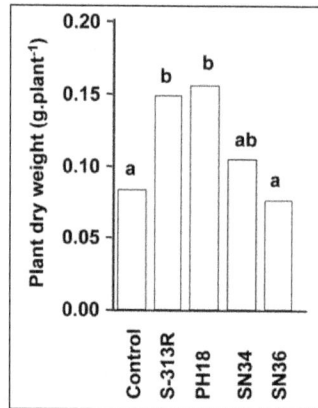

Plant growth promotion by mutant derivatives of Pseudomonas putida S-313R. Tomato plants were grown for 26 d in unsterilized compost either (a) without bacterial inoculation, (b) after inoculation with the wild-type strain S-313R, (c) after inoculation with an sftR mutant unable to use sulphate esters as S source in vitro (strain PH18), (d) after inoculation with an ssuE mutant unable to use aryl- or alkylsulphonates (strain SN34), or (e) after inoculation with an asfA mutant unable to use arylsulphonates (strain SN36). Plant dry weight was determined for five replicates. Letters show values that are statistically different from each other, determined using ANOVA (P=0.05) with Statview software.

References

- Scope-and-importance-of-soil-microbiology: agriinfo.in, Retrieved 18 April, 2019

- Microorganisms-found-in-soil-microbiology: biologydiscussion.com, Retrieved 14 June, 2019

- The-role-of-soil-microorganisms: sesl.com.au, Retrieved 28 February, 2019

- Microorg: compost.css.cornell.edu, Retrieved 19 May, 2019

- Microbial-biomass: soilquality.org.au, Retrieved 28 April, 2019

- Organic-matter-and-microorganisms-in-soil, soil-microbiology: biologydiscussion.com, Retrieved 1 August, 2019

- The-carbon-cycle, diagram, soil-microbiology: biologydiscussion.com, Retrieved 13 July, 2019

- The-phosphorus-cycle, diagram, soil-microbiology: biologydiscussion.com, Retrieved 29 April, 2019

- Factors-affecting-microbial-community, soil-microbiology: biologydiscussion.com, Retrieved 17 March, 2019

Bacteria and Archaea in Soil

There are several categories of microbes which are present in the soil such as bacteria, actinomycetes, algae, protozoa and fungi. The most important function of bacteria and archaea is to fix nitrogen in the soil. The chapter closely examines the different types of bacteria in soil such as actinobacteria and rhizobia to provide an extensive understanding of the subject.

Soil Bacteria

Bacteria are a major class of microorganisms that keep soils healthy and productive. Bacteria are tiny one-celled organisms generally 4/100,000 of an inch wide (1 μm). A teaspoon of productive soil generally contains between 100 million and 1 billion bacteria. That is as much mass as two cows per acre. A ton of microscopic bacteria may be active in each acre." While bacteria may be small, they make up both the largest number and biomass (weight) of any soil microorganism. Figure shows ciliate protozoa consuming bacteria.

Bacteria are similar in size to clay soil particles (<.2 μm) and silt soil particles (2-50 μm). They grow and live in thin water films around soil particles and near roots in an area called the rhizosphere. Bacteria's small size enables them to grow and adapt more rapidly to changing environmental conditions than larger, more complex microorganisms like fungi.

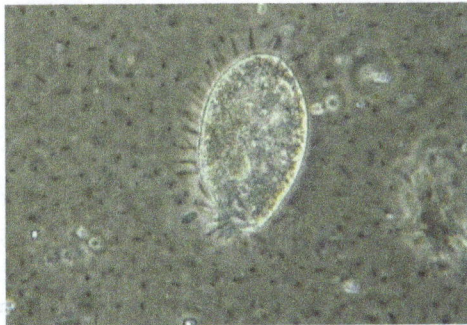

Close up view of a ciliate (protozoa) with various bacteria in the background.

Most soils are simply a graveyard for dead bacteria cells. Bacteria are so simple in structure that they have often been called a bag of enzymes and/or soluble bags of fertilizer. Since bacteria live under starvation conditions or soil water stress, they reproduce quickly when optimal water, food, and environmental conditions occur. Bacteria population may easily double in 15-30 minutes. Flourishing microbial populations increase soil productivity and crop yields over time.

Bacteria Classification

Most bacteria are classified into one of the following four categories:

Bacteria Shape

When scientists started first classifying bacteria, they started by looking at their basic shape. Bacteria generally have three major shapes: rod, sphere or spiral. Actinomycetes are still classified as bacteria but are similar to fungi except they are smaller in size. Classifying bacteria by shape is complex because many bacteria have different shapes and different arrangements.

Aerobic and Anaerobic Bacteria

Most microbes are generally inactive and may only have short burst of soil activity. Soil oxygen levels often determine soil bacteria activity. Most soil bacteria prefer well-oxygenated soils and are called aerobic bacteria and use the oxygen to decompose most carbon compounds. Examples of aerobic bacteria include the Aerobacter genus which is widely distributed in the soil and actinomycetes bacteria genus Streptomyces which give soil its good "earthy" smell.

Anaerobic bacteria prefer and some require an environment without oxygen. Anaerobic bacteria are generally found in compacted soil, deep inside soil particles (microsites), and hydric soils where oxygen is limiting. Many pathogenic bacteria prefer anaerobic soil conditions and are known to outcompete or kill off aerobic bacteria in the soil. Many anaerobic bacteria are found in the intestines of animals and are associated with manure and bad smells.

Gram Negative and Gram Positive Bacteria

When a staining agent is used in the lab, bacteria can be classified as "gram negative" or "gram positive." The staining agent attaches to the bacteria's cell walls. Gram negative bacteria are generally the smallest bacteria and are sensitive to drought and water stress. Gram positive bacteria are much larger in size, have thicker cell walls, negative charges on the outside cell wall surface and tend to resist water stress. Bacteroides are anaerobic gram negative bacteria that live in the gut of humans and animals. Listeria is a gram positive aerobic rod shaped bacteria found in contaminated food.

Other Bacteria Classifications

Another way to classify bacteria is by their growth and reproduction. Autotrophic bacteria (also called autotrophs) process carbon dioxide to get their carbon. Some autotrophic bacteria directly use sunlight and carbon dioxide to produce sugars, while others depend on other chemical reactions to obtain energy. Algae and cyanobacteria are some examples of autotrophic bacteria. Heterotrophic bacteria obtain their carbohydrates and/or sugars from their environment or the living organism or cell they inhabit. Examples include Arthrobacter bacteria involved in nitrogen nitrification.

With new advances in DNA sequencing, most scientists are classifying bacteria based on the type of environment in which they inhabit. Bacteria can live in extreme environments like hot springs for sulfur bacteria or in extreme cold as in ice water in the Arctic. Bacteria may also be classified by living in a highly acidic versus alkaline environment, aerobic versus anaerobic, or autotrophic versus heterotrophic environment.

Bacteria Functional Groups

Bacteria perform many important ecosystem services in the soil including improved soil structure and soil aggregation, recycling of soil nutrients, and water recycling. Soil bacteria form microaggregates in the soil by binding soil particles together with their secretions. These microaggregates are like the building blocks for improving soil structure. Improved soil structure increases water infiltration and increases water holding capacity of the soil.

Bacteria perform important functions in the soil, decomposing organic residues from enzymes released into the soil. Ingham describes the four major soil bacteria functional groups as decomposers, mutualists, pathogens and lithotrophs. Each functional bacteria group plays a role in recycling soil nutrients.

The decomposers consume the easy-to-digest carbon compounds and simple sugars and tie up soluble nutrients like nitrogen in their cell membranes. Bacteria dominate in tilled soils but they are only 20-30 percent efficient at recycling carbon (C). Bacteria are higher in nitrogen (N) content (10-30 percent nitrogen, 3 to 10 C:N ratio) than most microbes.

Of the mutualistic bacteria, there are four bacteria types that convert atmospheric nitrogen (N_2) into nitrogen for plants. There are three types of soil bacteria that fix nitrogen without a plant host and live freely in the soil and these include Azotobacter, Azospirillum and Clostridium.

The Rhizobium bacteria (gram negative rod-shaped bacteria) species associate with a plant host: legume (alfalfa, soybeans) or clover (red, sweet, white, crimson) to form nitrogen nodules to fix nitrogen for plant growth. The plant supplies the carbon to the Rhizobium in the form of simple sugars. Rhizobium bacteria take nitrogen from the atmosphere and convert it to a form the plant can use. For plant use, the atmospheric nitrogen (N_2) or reactive nitrogen combines with oxygen to form nitrate (NO_3^-) or nitrite (NO_2^-) or combines with hydrogen to produce ammonia (NH_3^+) or ammonium (NH_4^+) which are used by plant cells to make amino acids and proteins.

Nitrogen fixing Rhizobium bacteria form nodules on a soybean root

Many soil bacteria process nitrogen in organic substrates, but only nitrogen fixing bacteria can process the nitrogen in the atmosphere into a form (fixed nitrogen) that plants can use. Nitrogen fixation occurs because these specific bacteria produce the nitrogenase enzyme. Nitrogen fixing bacteria are generally widely available in most soil types (both free living soil species and bacteria species dependent on a plant host). Free living species generally only comprise a very small percentage of the total microbial population and are often bacteria strains with low nitrogen fixing ability.

Nitrification is a process where nitrifying bacteria convert ammonia (NH_4^+) to nitrite (NO_2^-) and then to nitrate (NO_3^-). Bacteria and fungi are typically consumed by protozoa and nematodes and the microbial wastes they excrete is ammonia (NH_4^+) which is plant available nitrogen. Nitrite bacteria (Nitrosomonas spp.) convert the ammonia into nitrites (NO_2^-) and nitrate bacteria (Nitrobacter spp.) may then convert the nitrites (NO_2^-) to nitrates (NO_3^-). Nitrifying bacteria prefer alkaline soil conditions or a pH above 7. Both nitrate and ammonia are plant available forms of nitrogen; however, most plants prefer ammonia because the nitrate has to be converted to ammonia in the plant cell in order to form amino acids.

Denitrifying bacteria allow nitrate (NO_3^-) to be converted to nitrous oxide (N_2O) or dinitrogen (N_2) (atmospheric nitrogen). For denitrification to occur, a lack of oxygen or anaerobic conditions must occur to allow the bacteria to cleave off the oxygen. These conditions are common in ponded or saturated fields, compacted fields, or deep inside the microaggregates of soil where oxygen is limited. Denitrifying bacteria decrease the nitrogen fertility of soils by allowing the nitrogen to escape back into the atmosphere. On a saturated clay soil, as much as 40 to 60 percent of the soil nitrogen may be lost by denitrification to the atmosphere.

Pathogenic bacteria cause diseases in plants and a good example are bacteria blights. Healthy and diverse soil bacteria populations produce antibiotics that protect the plants from disease causing organisms and plant pathogens. Diverse bacteria populations compete for the same soil nutrients and water and tend to act as a check and balance system by reducing the disease-causing organism populations. With high microbial diversity, soils have more nonpathogenic bacteria competing with the pathogenic bacteria for nutrients and habitat. Streptomycetes (actinomycetes) produce more than 50 different antibiotics to protect plants from pathogenic bacteria.

Lithotrophs (chemoautotrophs) get their energy from compounds other than carbon (like nitrogen or sulfur) and include species important in nitrogen and sulfur recycling. Under well-aerated conditions, sulfur-oxidizing bacteria make the sulfur more plant available while under saturated (anaerobic, low oxygen) soil conditions, sulfur reducing bacteria make sulfur less plant available.

Actinomycetes have large filaments or hyphae and act similar to fungus in processing soil organic residues which are hard to decompose (chitin, lignin, etc.). When farmers plow or till the soil, actinomycetes release "geosmin" as they die which gives freshly turned soil its characteristic smell. Actinomycetes decompose many substances but are more active at high soil pH levels. Actinomycetes are important in forming stable humus, which enhances soil structure, improves nutrient storage, and increases water retention.

Soil Benefits from Bacteria

Bacteria grow in many different microenvironments and specific niches in the soil. Bacteria populations expand rapidly and the bacteria are more competitive when easily digestible simple sugars are readily available around in the rhizosphere. Root exudates, dead plant debris, simple sugars, and complex polysaccharides are abundant is this region. About 10 to 30 percent of the soil microorganisms in the rhizosphere are actinomycetes, depending on environmental conditions.

Many bacteria produce a layer of polysaccharides or glycoproteins that coats the surface of soil particles. These substances play an important role in cementing sand, silt and clay soil particles into stable microaggregates that improve soil structure. Bacteria live around the edges of soil mineral particles, especially clay and associated organic residues. Bacteria are important in producing polysaccharides that cement sand, silt and clay particles together to form microaggregates and improve soil structure. Bacteria do not move very far in the soil, so most movement is associated with water, growing roots or hitching a ride with other soil fauna like earthworms, ants, spiders, etc.

In general, most soil bacteria do better in neutral pH soils that are well oxygenated. Bacteria provide large quantities of nitrogen to plants and nitrogen is often lacking in the soil. Many bacteria secrete enzymes in the soil to makes phosphorus more soluble and plant available. In general, bacteria tend to dominate fungi in tilled or disrupted soils because the fungi prefer more acidic environments without soil disturbance. Bacteria also dominate in flooded fields because most fungi do not survive without oxygen. Bacteria can survive in dry or flooded conditions due to their small size, high numbers, and their ability to live in small microsites within the soil where environmental conditions may be favorable. Once the environmental conditions around these micro-sites become more favorable, the survivors quickly expand their populations. Protozoa tend to be the biggest predators of bacteria in tilled soils.

In order for bacteria to survive in the soil, they must adapt to many microenviron-ments. In the soil, oxygen concentrations vary widely from one microsite to another. Large pore spaces filled with air provide high levels of oxygen, which favors aerobic conditions, while a few millimeters away, smaller micropores may be anaerobic or lack oxygen. This diversity in soil microenvironments allows bacteria to thrive under var-ious soil moisture and oxygen levels, because even after a flood (saturated soil, lack of oxygen) or soil tillage (infusion of oxygen) small microenvironments exist where different types of bacteria and microorganisms may live to repopulate the soil when environmental conditions improve.

Natural succession happens in a number of plant environments including in the soil. Bacteria improve the soil so that new plants can become established. Without bacte-ria, new plant populations and communities struggle to survive or even exist. Bacte-ria change the soil environment so that certain plant species can exist and proliferate. Where new soil is forming, certain photosynthetic bacteria start to colonize the soil, recycling nitrogen, carbon, phosphorus, and other soil nutrients to produce the first organic matter. A soil that is dominated by bacteria usually is tilled or disrupted and has higher soil pH and nitrogen available as nitrate, which is the perfect environment for low successional plants called weeds.

As the soil is disturbed less and plant diversity increases, the soil food web becomes more balanced and diverse, making soil nutrients more available in an environment better suited to higher plants. Diverse microbial populations with fungus, protozoa and nematodes keep nutrients recycling and keep disease-causing organisms in check.

Actinobacteria

Actinobacteria are a group of Gram-positive bacteria with high guanine and cytosine content in their DNA, which can be terrestrial or aquatic. Though they are unicellular like bacteria, they do not have distinct cell wall, but they produce a mycelium that is non-septate and more slender. Actinobacteria include some of the most common soil, fresh-water, and marine type, playing an important role in decomposition of organic materials, such as cellulose and chitin, thereby playing a vital part in organic matter turnover and carbon cycle, replenishing the supply of nutrients in the soil, and is an important part of humus formation. Actinobacterial colonies show powdery consistency and stick firmly to agar surface, producing hyphae and conidia/sporangia-like fungi in culture media.

Actinobacteria produce a variety of secondary metabolites with high pharmacological and commercial interest. With the discovery of actinomycin, a number of antibiotics have been discovered from Actinobacteria, especially from the genus Streptomyces. They are widely distributed in soil with high sensitivity to acid and low pH. Actinobacteria have a number of important functions, including degradation/decomposition of all sorts of organic substances such as cellulose, polysaccharides, protein fats, organic acids, and so on. They are also responsible for subsequent decomposition of humus (resistant material) in soil and for the earthy smell of freshly ploughed soils, producing a number of antibiotics like streptomycin, terramycin, aureomycin, and so on.

Habitat of Actinobacteria

Terrestrial Environment

Soil remains the most important habitat for Actinobacteria with streptomycetes existing as a major component of its population. According to numerous reports, Streptomyces was encountered to be the most abundant genus isolated in each of the study. Terrestrial Actinobacteria have various interesting antimicrobial potentials. Oskay et al isolated Actinobacteria that had capability of producing novel antibiotics with high antibacterial activity. In anoxic mangrove rhizosphere, Actinobacterial species such as Streptomyces, Micromonospora, and Nocardioform were found to be abundant, which is 1000 to 10000 times smaller than arable lands because of tidal influence. Similarly, Nocardia isolated from mangrove soil produced new cytotoxic metabolites that strongly inhibited human cell lines, such as gastric adenocarcinoma. Dessert soil is also considered as an extreme terrestrial environment where only certain species, especially wherein Actinobacteria, often use Microcoleus as a source of food. There are several reports showing the distribution of Actinobacteria in various locations, such as sandy soil (Cario, Egypt; Falmouth, MA), black alkaline soil (Karnataka, India), sandy loam soil (Keffi Metropolis, Nigeria; Presque Isle, PA), alkaline dessert soil (Wadi El Natrun, Egypt; Wadi Araba, Egypt), and subtropical dessert soil (Thar, Rajasthan), where Streptomyces sp. were dominant followed by the other organisms, such as Nocardia, Nocardiopsis, and Actinomycetes. In the study of Nithya et al, 134 morphologically distinguished culturable Actinobacteria were isolated from 10 different desert soil samples, and the isolates were found to have varying level of antibacterial activity against bacterial pathogens. Equally, Actinobacteria play a major part in rhizosphere microbial community in the turnover of recalcitrant plant organic matter, and thus the rhizosphere region is considered as one of the best habitats for isolation of these microorganisms. Priyadharsini et al in her study isolated 45 morphologically distinct colonies from 12 different paddy field soils and observed their ability to inhibit the growth of Cyperus rotundus. The isolates include Streptomyces sp., Streptoverticillium sp., Actinomadura sp., Kitasatosporia sp., Nocardiopsis sp., Pseudonocardia sp., and Kibdelosporangium sp.

General Characteristics of Actinobacteria

Actinobacteria comprises a group of branching unicellular microorganisms, most of which are aerobic-forming mycelium known as substrate and aerial. They reproduce by binary fission or by producing spores or conidia, and sporulation of Actinobacteria is through fragmentation and segmentation or conidia formation. The morphological appearance of Actinobacteria is compact, often leathery, giving a conical appearance with a dry surface on culture media and are frequently covered with aerial mycelium.

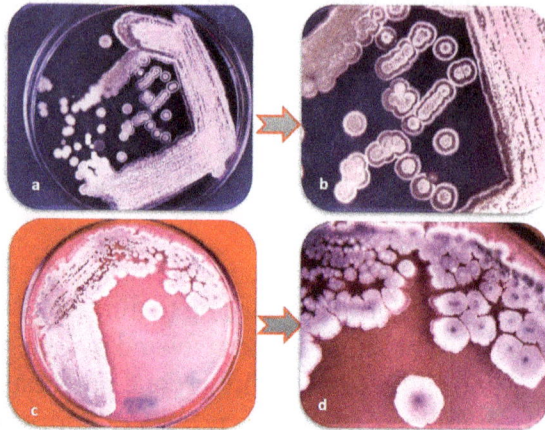

Appearance of Actinobacteria isolates on Starch casein agar plate. a, c Plate view of the Actinobacterial isolates. b, d Morphology of individual colonies.

Aerial Mycelium

The aerial mycelium is usually thicker than the substrate mycelium. The aerial mycelium shows sufficient differentiation that a miscellaneous assortment of isolates can be segregated into a number of groups having similar morphological characteristics under fixed condition. This is designated as one of the most important criteria for the classification of the genus Streptomyces into species, comprising structure (cottony, velvety, or powdery), formation of rings or concentric zones, and pigmentation.

Substrate Mycelium

The substrate mycelium of Actinobacteria varies in size, shape, and thickness. Its color ranges from white or virtually colorless to yellow, brown, red, pink, orange, green, or black.

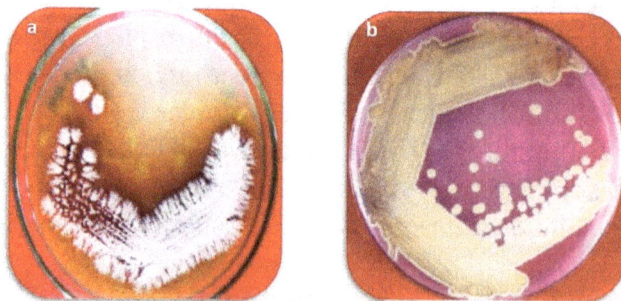

Abundant growth of Actinobacterial isolate on starch casein agar medium. a. Aerial mycelium. b. Reverse side of plate showing substrate mycelium.

Morphological Appearance

Morphology has been an important characteristic to identify Actinobacteria isolates, which was used in the first descriptions of Streptomyces species. This is made using various standard culture media, including International Streptomyces Project (ISP). Various morphological observations, including germination of spores, elongation and branching of vegetative mycelium, formation of aerial mycelium, color of aerial and substrate mycelium, and pigment production, have been used to identify Actinobacteria. Light microscopy was used to study the formation of aerial mycelium and substrate mycelium, and scanning electron microscopy was used to study the spores, the spore surface, and spore structure.

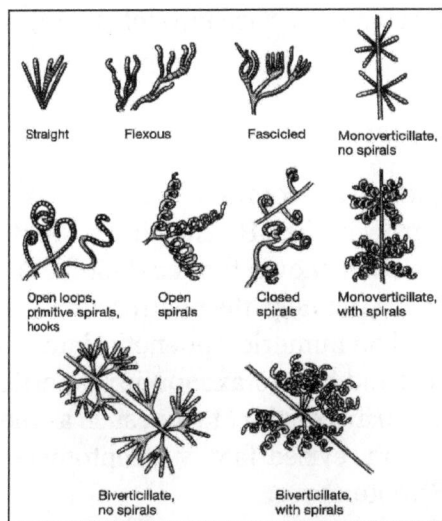

Type of spore-bearing structure in streptomycetes.

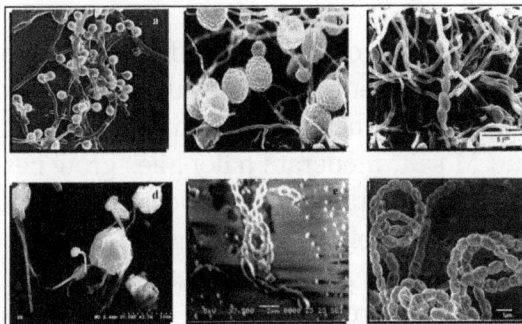

Scanning electron photographs of various Actinobacterial isolates. a. Micromonospora sp. b. Streptosporangium sp. c. Saccharopolyspora sp. d. Actinosynnema sp. e. S. noursei DPTD21. f. Streptomyces sp. JD9.

Types of Actinobacteria

Thermophilic Actinobacteria

Number of studies has been carried out by the researchers to confirm the existence of extremophilic and extreme tolerant soil Actinobacteria (acid tolerant and alkali tolerant, psychrotolerant and thermotolerant, and halotolerant and haloalkalitolerant or xerophilic). Mesophilic Actinobacteria can grow at an optimal temperature from 20 °C to 42 °C, among which thermotolerant species exist, which can survive at 50 °C. Moderately thermophilic Actinobacteria have an optimum growth at 45 °C–55 °C, whereas strictly thermophilic Actinobacteria grow at 37 °C–65 °C with the optimum temperature at 55 °C–60 °C. Incubation temperatures of 28 °C, 37 °C, and 45 °C are considered optimal for isolation of soil mesophilic, thermotolerant, and moderately thermophilic Actinobacteria. Thermoactinomyces, which is presently excluded from the order Actinomycetales, are described as thermophilic forms depending on its phenotypic and molecular genetic characteristics, as well as among some species of Thermomonospora, Microbispora, Saccharopolyspora, Saccharomonospora, and Streptomyces.

Acidophilic Actinobacteria

Acidophilic Actinobacteria, which are common in terrestrial habitats such as acidic forest and mine drainage soil, grow in the pH range from about 3.5 to 6.5, with optimum rates at pH 4.5 to 5.5. It has been shown that acidophilic Actinobacteria consistently form two distinct aggregate taxa (namely, the neutrotolerant acidophilic and strictly acidophilic cluster groups) based on numerical phenetic data; members of the two groups share common morphological and chemotaxonomic properties. Also some members of the strictly acidophilic group form a distinct taxon, such as the genus Streptacidiphilus, which has been assigned to the revised family Streptomycetaceae, together with the genera Kitasatospora and Streptomyces.

Halophilic Actinobacteria

Halophilic Actinobacteria are categorized into different types based on their growth in media containing different concentrations of salt. Extreme halophiles grow best in media containing 2.5–5.2 M salt, whereas borderline extreme halophiles grow best in media containing 1.5–4.0 M salt, moderate halophiles grow best in media containing 0.5–2.5 M salt, and finally halotolerants that do not show an absolute requirement to salt for growth but grow well up to often very high salt concentrations and tolerate 100 g/l salt (equivalent to 1.7 M NaCl) at least. Seawater, saline soils, salt lakes, brines, and alkaline saline habitats are considered as the best habitats for isolating halophilic Actinobacteria. Generally, most of the halophilic Actinobacteria have been isolated from saline soils. Halophilic Actinobacteria isolated from marine environments are assigned to a few genera, including Micromonospora, Rhodococcus, and Streptomyces. The other group includes Dietzia, Salinispora, Marinophilus, Solwaraspora, Salinibacterium, Aeromicrobium, Gordonia, Microbacterium, Mycobacterium, Nocardiopsis, Pseudonocardia, Actinomadura, Saccharopolyspora, Streptosporangium, Nonomuraea, Williamsia, and Verrucosispora.

Bioremediation

Actinobacteria possess many properties that make them good candidates for application in bioremediation of soils contaminated with organic pollutants. In some contaminated sites, Actinobacteria represent the dominant group among the degraders. They play an important role in the recycling of organic carbon and are able to degrade complex polymers. Sanscartier et al reported that the greater use of petroleum hydrocarbons that are widely used in our daily life as chemical compounds and fuel has become one of the most common contaminants of large soil surfaces and eventually is considered as a major environmental problem. Some reports suggests that Streptomyces flora could play a very important role in degradation of hydrocarbons. Many Actinobacterial strains have the ability to solubilize lignin and degrade lignin-related compounds by producing cellulose- and hemicellulose-degrading enzymes and extracellular peroxidase. Actinobacteria species have the ability to live in an oily environment and thus they can be used in bioremediation to reduce oil pollutants. Nocradiopsis sp. SD5 degraded feather waste by producing keratinase enzyme.

Control of Plant Diseases

The worldwide efforts in the search of natural products for the crop protection market have progressed significantly, and Actinobacteria, especially genus Streptomyces, appear to be good candidates in finding new approaches to control plant diseases. The agroindustry shows a marked interest for Actinobacteria as a source of agroactive compounds of plant growth–promoting rhizobacteria (PGPR) and of biocontrol tools. About 60% of the new insecticides and herbicides reported in the past 5 years originate from Streptomyces. Kasugamycin is a bactericidal and fungicidal metabolite discovered in Streptomyces kasugaensis, which acts as an inhibitor of protein biosynthesis in microorganisms but not in mammals, and its toxicological properties are excellent. To market the systemically active kasugamycin for control of rice blast Pyricularia oryzae and bacterial Pseudomonas diseases in several crops, Hokko Chemical Industries developed a production process. Polyoxin B and D were isolated as metabolites of Streptomyces cacaoi var. asoensis in 1965 by Isono et al as a new class of natural fungicides. The ability of the polyoxins to interfere with the fungal cell wall synthesis by specifically inhibiting chitin synthase makes them acceptable with regard to environmental considerations. Polyoxin B found application against a number of fungal pathogens in fruits, vegetables, and ornamentals. Polyoxin D is marketed by several companies to control rice sheath blight caused by Rhizoctonia solani. The validamycin family was detected by Takeda researchers in 1968 in a greenhouse assay when screening streptomycete extracts for activity against rice sheath blight. Validamycin A was found to be a prodrug, which is converted within the fungal cell to validoxylamine A, an extremely strong inhibitor of trehalase. This mode of action gives validamycin A a favorable biological selectivity because vertebrates do not depend on the hydrolysis of the disaccharide trehalose for their metabolism. Inhibition of plant pathogenic Rhizoctonia solani under in vitro condition was assessed with the culture supernatant of Streptomyces sp., which showed that the tested Actinobacteria had the ability to reduce damping off severity in tomato plants. Table represents some of the antibiotics produced by the Actinobacteria that suppresses various plant diseases.

Disease	Actinobacteria	Antibiotic produced
Potato scab	Streptomyces melanosporofaciens EF-76 and FP-54	Geldanamycin
Grass seedling disease	Streptomyces violaceusniger YCED9	Nigericin and guanidylfungin A
Root rot of Pea	Streptomyces hygroscopicus var. geldanus	Geldanamycin
Asparagus root diseases	Streptomyces griseus	Faeriefungin
Rice blast disease	Streptomyces kasugaensis	Kasugamycin
Broad range of plant diseases	Streptomyces griseochromogenes	Blasticidin S
Sheath blight of rice	Streptomyces hygroscopicus var. limoneus No. T-7545	Validamycin
Brown rust of wheat	Streptomyces hygroscopicus	Gopalamycin
Phytophthora blight of pepper	Streptomyces violaceusniger	Tubercidin
Phytophthora blight of pepper	Streptomyces humidus	Phenylacetic Acid
Damping-off of cabbage	Streptomyces padanus	Fungichromin
Rice sheath blight	Streptomyces cacaoi var. asoensis	Polyoxin B and D
Powdery mildew	Streptoverticillium rimofaciens	Mildiomycin
Rice root disease	Micromonospora sp. SF-1917	Dapiramicin
Rice blast	Micromonospora sp. M39	2,3-dihydroxybenzoic acid, phenylacetic acid, cervinomycin A1 and A2
Blotch of wheat	Streptomyces malaysiensis	Malayamycin
Powdery mildew of cucumber	Streptomyces sp. KNF2047	Neopeptin A and B

Plant disease suppression by antibiotics produced by Actinobacteria.

Nematode Control

It has been known for decades that effective control of plant-parasitic nematodes is dependent on chemical nematicides. Due to its ill effects with respect to the environmental hazards, hazardous nematicides have emphasized the need for new methods to control nematodes. Today, numerous microorganisms are recognized as antagonists of plant-parasitic nematodes. Especially, Actinobacteria have potential for use in biological control as they are known to produce antibiotics. The production of avermectins by a species of Streptomyces shows that soil-borne organisms can produce highly nematicidal compounds. S. avermitilis produces ivermectin, which has an excellent activity against Wucheria bancroftii. Similarly various other antiparasitic compounds are produced from different Streptomyces sp., Salinispora sp., and Marinactinospora sp., which includes Milbemycin, Antimycin A9, Fervenulin, bafilolides, Valinomycin, Salinosporamide A, Kalafungin, Thiamycins, and Axenomycins.

Enhancement of Plant Growth

Despite the well-documented history of Streptomyces in biocontrol and preliminary evidence of their capacity to enhance plant growth, Streptomyces species have been poorly investigated specifically for their potential as PGPR. While the beneficial effect of some strains of PGPR on particular crops is certain, the mechanisms employed by PGPR are unclear. PGPR can affect plant growth in two general ways, either directly or indirectly. Indirect promotion occurs when PGPR lessen or prevent the harmful effects of one or more deleterious microorganisms. This is chiefly attained through biocontrol or the antagonism of soil plant pathogens. Specifically, colonization or the biosynthesis of antibiotics and other secondary metabolites can prevent pathogen invasion and establishment. Direct promotion of plant growth by PGPR occurs when the plant is supplied with a compound that is synthesized by the bacteria, or when PGPR otherwise facilitates plant uptake of soil nutrients. Merriman et al reported the use of S. griseus for seed treatment of barley, oat, wheat, and carrot to increase their growth.

The isolate was originally selected for the biological control of Rhizoctonia solani. Though the S. griseus isolate did increase the average grain yield, dry foliage weight, tiller number, and advanced head emergence for both wheat and oat over controls, the differences were not statistically significant. As a seed treatment for carrot, the isolate was more successful. Marketable yields were increased over controls by 17% and 15% in two separate field trials. Specifically, both trials also indicated an increased yield of large and very large grade carrots over controls. El-Abyad et al described the use of three Streptomyces spp. in the control of bacterial, Fusarium and Verticillium wilts, early blight, and bacterial canker of tomato. The isolates used were Streptomyces pulcher, Streptomyces canescens, and Streptomyces citreofluorescens. In addition, tomato growth was observed to be significantly improved with the antagonistic Streptomyces spp. as a seed coating. An increased availability of growth regulators produced by the inoculum was the reason proposed for the improvement in tomato growth, although this was not formally tested. The information available on streptomycetes as plant growth promoters is limited, so is the information describing the possibility of their direct growth promotion mechanisms. Like most rhizobacteria, it seems highly probable that streptomycetes are capable of directly enhancing plant growth.

Rhizobia

Rhizobium are a group of Gram-negative soil bacteria that are well known for their symbiotic relationship with various leguminous (soybeans, alfalfa etc).

There are different types of Rhizobium that are categorized on the basis of the rate of growth and the type of plant they are associated with.

Some species of Rhizobium include:

- R. leguminosarum
- R. alamii
- R. lentis
- R. japonicum
- R. metallidurans
- R. smilacinae
- R. phaseoli
- R. trifolii

Classification

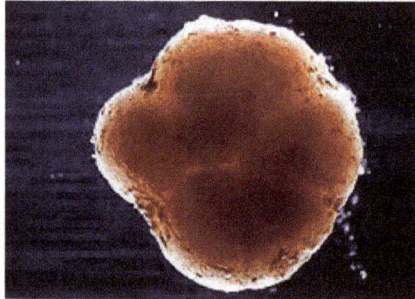

Along with Bradyrhizobium, Sinorhizobium, Mesorhizobium, Azorhizobium, and Allorhizobium, Rhizobium is a soil Rhizobia, which means that it consists of bacteria with the ability to fix nitrogen. As such, it presents a significant advantage to the plants it infects by contributing to their growth and development.

It is classified as follows:

- Kingdom: Bacteria - Like other bacteria, Rhizobium cells lack membrane-bound organelles. They also have ribosome, cytoplasm and flagella.
- Phylum: Proteobacteria - Phylum proteobacteria includes Gram-negative bacteria including those responsible for fixing nitrogen.
- Class: Alphaproteobacteria - Class Alphaproteobacteria is composed of a wide range of Gram-negative with varying life characteristics.
- Order: Rhizobiales - This is an order of Gram-negative Alphaproteobacteria that form a symbiotic relationship with plant roots (where they fix nitrogen).

- Family: Rhizobiaceae - Along with such genera as Agrobacterium, genus Rhizobium makes up the family Rhizobiaceae. While some in this family have been shown to negatively affect plant development, Rhizobium (a member of Rhizobia) contribute to plant nutrition through nitrogen-fixation.

- Species: Includes such species as Rhizobium leguminosarum and Rhizobium lentil.

In addition to this classification, Rhizobium bacteria are also categorized based on the species of legume that they nodulate. This type of grouping is known as cross-inoculation.

Based on studies on a wide variety of legumes, it became evident that not all Rhizobia are capable of nodulating all types of legumes. This resulted in a need to group Rhizobium into appropriate groups (cross-inoculation groups) which are simply groups of legumes that given species of Rhizobium nodulate.

The cross-inoculation groups include:

- Clover groups - R. trifolii infects and nodulates plants of genus Trifolium (clovers/trefoil).

- Alfalfa groups - R. meliloti infects and nodulates the roots of medicago, melilotus and medicago.

- Bean group - R. phaseoli infects and nodulates plants of genus Phaseolus (e.g. beans).

- Lupine group - R. lupine nodulates lupines and serradella (Ornithopus).

- Pea group - R. leguminosarum infects and nodulates pea, sweet pea, lentil, and vetch.

- Soybean group - R. japonicum nodulates Glycine such as soybean.

- Cowpea group - Rhizobium sp. nodulates cowpea, pegionpea, lespedza, groundnut and kudz among a few others.

Characteristics (Rhizobium Leguminosarum)

Discovered and described in 1889, R. leguminosarum is the type species of Rhizobium in the same way Rhizobium has been the type genus of family Rhizobiaceae. As such, it can be viewed as a representative of the genus (Rhizobium).

Some of the characteristics of the bacteria include:

- They appear as elongated rods when viewed under the microscope.

- Like a number of other bacteria, Rhizobium leguminosarum do not form spores in their life cycle.

- They posses several flagella on their polar end. This allows them to move from one location to another.

- They are aerobic. As such, they need oxygen for respiratory purposes.

- There are various strains of the bacteria some of which have granules.

- They are Gram-negative bacteria.

- Although they can tolerate higher temperatures of about 38 degrees Celsius, Rhizobium leguminosarum ideally grow in temperatures of between 20 and 28 degrees Celsius.

- Apart from various types of carbohydrates, the bacteria also uses nitrates and nitrite, ammonium salts and various amino acids among others for development.

There are different strains which include:

- Rhizobium leguminosarum biovar viciae 3841.

- Rhizobium leguminosarum USDA2370T.

- Rhizobium pisi DSM30132T.

- Rhizobium fabae CCBAU33202T.

Infection (Penetration)

Rhizobium species like R. leguminosarum can be found in soil. However, the root of leguminous plants (lentil, sweetpea etc) is their primary habitat. In the soil, various leguminous plants release various exudates (dicarboxylic acids etc) that attract Rhizobium species.

Flavanoids have also been shown to play an important role attracting the bacteria given that they are easily absorbed through the membrane of the organisms (passively). Once the bacteria detect these chemicals, they actively swim towards and attach to the legume root.

In addition to attracting the bacteria, these chemicals (flavanoids in particular) also play an important role of activating genes involved in producing Nod factors. Here, then, attraction to the legume roots is followed by transcription of Nod genes in preparation of the symbiotic relationship.

For the plant, Nod factors stimulate the branching of root hair, hydrolysis of the cell wall as well as deformation of the cell wall. Having attracted the bacteria through the exudates, the changes in the plant roots make it easy for the organism to enter the cells of the root hair for symbiosis.

When the bacteria comes into contact with the root hair, they cause the plasma membrane of the cells to invaginate. As the bacteria penetrates the cell, the plant produces new cell wall material at the site to not only cover the bacteria, but also allowing them to enter deeper into the root hairs.

Once the bacteria infects the cells of the root hair, the symbiotic process may produce the following types of nodules:

- Determinate nodules - Determinate are found in plants like soybeans, are spherical in shape with large lenticels. Compared to indeterminate nodules, determinate nodules are formed once the tip of the root hairs start growing. However, they lack the persistent meristem found in indeterminate nodules.

- Indeterminate nodules - Compared to determinate nodules, indeterminate nodules are cylindrical in shape and frequently branched. They are often found in plants like peas and alfalfa and start developing even before the growth of the root tip.

Whereas indeterminate nodules are found in temperate regions, determinate nodules are commonly formed in tropical and sub-tropical areas.

Determinate nodules produce ureide products while indeterminate nodules amide products.

Nitrogen Fixation

Nitrogen fixation in the nodules begins when the nodules fully mature. Here, it is worth noting that nitrogen fixation involves the conversion of atmospheric nitrogen into organic compounds (particularly ammonia) that can be used for plant development.

This process requires two important genes (nif and fix). These genes play an important role of producing several crucial enzymes that are involved in the nitrogen fixation.

The process requires the following:

- Enzymes (dinitrogenase and dinitrogenase reductase),
- ATP energy,
- Atmospheric Nitrogen (Nitrogen molecules).

During nitrogen fixation, the enzyme nitrogenase is involved in the breaking of the bonds that hold Nitrogen atoms together (covalent bonds). In their atmospheric state, nitrogen molecules are non-reactive given that they are bound by covalent bonds. By breaking this bond, the Nitrogen atoms are free to form bonds with other atoms.

This process (breaking down the covalent bonds) requires a lot of energy. Given that Rhizobium bacteria are not capable of making their own food for energy, they rely on the plant (in the rhizospere) to provide sources of energy.

By using energy sources from the plant, the bacteria gains sufficient energy that makes it possible for the enzymes to break down Nitrogen molecules into Nitrogen atoms.

Nitrogenase enzyme is oxygen sensitive. However, high metabolic activity of the bacteria as well as a diffusion barrier developed at the nodule periphery help protect the enzyme from the high level of oxygen.

Nitrogen fixation in the nodules takes place while the Nitrogen molecules are attached to the enzyme. While the nitrogen is bound to the enzyme, electrons provided by ferredoxin make it possible for the Iron protein of the enzyme to be reduced. This protein then binds to the ATP, which in turn causes molybdenum-iron protein (also a component of the enzyme) to be reduced.

This reduction provides electrons required to break down the Nitrogen molecules and produce a compound known as Diimide ((NH)2). The process is repeated two more times which further reduces the Diimide into two ammonia molecules.

The process is represented as follows:

$$N_2 + 8H + 8e^- + 16Mg\text{-}ATP\ 2NH_3 + H_2 + 16MgADP + 16Pi$$

The relationship between leguminous plants and Rhizobium bacteria is referred to as a symbiotic relationship because the bacteria and the plant benefit each other. While the plant's rhizosphere provides shelter and a source of energy for the bacteria, the bacteria converts atmospheric nitrogen to ammonia which is required for proper growth and development of the plant.

Biofertilizer

Nitrogen is one of the most important elements in nature given that it is used to make various products that plants require for their development. For instance, using such products as ammonia and nitrates that are made from nitrogen, plants are able to form the protein they need for their development.

While various chemical fertilizers are successfully used to increase yields, they are also expensive and tend to pollute the environment. Given that Rhizobium bacteria have a good and beneficial relationship with various leguminous plants, there has been increased interest to use them as biofertilizers.

In small scale scenarios, the inoculation of Rhizobium in various cereal grains has been shown to help increase yields. This was also shown to result in an increased uptake of nitrogen thus benefiting the plants.

On a large scale, production of the inoculants begins with identifying the most effective strain of bacteria through host-specificity. Here, research helps identify the most effective strain of Rhizobium that is then cultured for large scale production.

During mass culture production, the bacteria is cultured in large flasks composed of such carbohydrates as mannitol, arabinose and sucrose among others. This helps produce fertilizer products that have been shown to help enhance photosynthesis and thus yields such plants as banana and rice among others significantly.

Biofertilizer presents a significant advantage in that they not only increase yields for farmers, but are also safe and do not negatively affect the soil and environment they are used in. Studies in different parts of the world such as Sudan have found this method to be particularly beneficial for a variety of crops.

Culture

Sample Preparation:

- Carefully uproot a leguminous plant (sweetpeas, beans etc).

- Gently wash the roots using sterilized water to remove any soil, mud etc.

- Using a clean knife/blade, carefully remove/detach the nodules (which are visible to the naked eye) from the roots and place them in a container with clear water.

- After washing the nodules, place them in a container with a solution of merculic chloride for a few minutes - this helps disinfect and remove any microorganism on the surface of the nodule.

- Wash the nodules using alcohol and then wash with clean water several times.

- Place the nodules in a test-tube and add a few drops of water (distilled water).

- Using a glass rod, crush the nodules to release the bacteria into the water.

Culture Procedure:

- Gently pour mannitol-agar medium into 3 sterile test-tubes.

- Add a few drops of the sample (rhizobium water) into the test-tubes using sterilized droppers.

- Mix the contents thoroughly.

- Pour the contents into different Petri dishes and incubate at 45 degrees Celsius.

- Turn the Petri dish when the medium solidifies.

- Allow the culture to incubate for about 3 days.

Sample may also be collected from soil surrounding the roots of leguminous plant for comparison.

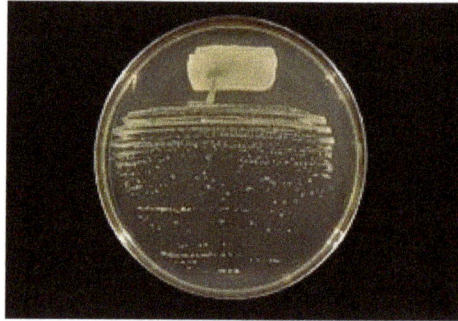

Gram Staining

Using a sample from the culture, Gram-staining involves the following steps:

- Fix the bacteria using methanol. Having placed the sample at the central part of a clean slide, add a few drops of methanol to fix. This helps preserve the morphology of the bacteria.

- Allow the slide to dry and pour the primary stain on the sample (crystal violet) - Allow to stand for a few minutes.

- Add a few drops of the mordant (Iodine solution).

- Decolorize the sample using ethanol/acetone gently.

- Pour safranin on the sample (counter-stain) for a few minutes.

- Wash gently with water to remove any excess counter-stain.

- Mount and view under the microscope.

Rhizobia-Legume Symbiosis

Ever since the identification by Hellriegel and Wilfarth of rhizobia as the source of fixed nitrogen in root nodules of legumes, people have wondered whether or not plants outside the Fabaceae could be manipulated to associate with rhizobia. The development of nodules, the keystone of Hellriegel and Wilfarth's findings, has since become the "Holy Grail" of the field of biological nitrogen fixation. It is well known that the rhizobia-legume interaction falls into cross inoculation groups, whereby certain rhizobial strains nodulate only certain legumes. For example, Sinorhizobium meliloti effectively nodulates species of Medicago, Melilotus and Trigonella, whereas Rhizobium leguminosarum bv viciae induces nitrogen-fixing nodules on Pisum, Vicia, Lens and Lathyrus spp. Closely related to the pea (Pisum sativum) strain is R. leguminosarum bvtrifolii, which initiates nodules only on species of clover (Trifolium).

However, not all rhizobia strain-legume associations are this tight. For example, Rhizobium strain NGR234 nodulated 232 species of legumes from 112 genera tested and even nodulated the nonlegume Parasponia andersonii, a member of the elm family. On the opposite end of the spectrum, not all members of the legume family nodulate. Of the three different subfamilies of legumes—Caesalpinoideae, Mimosoideae, and Papilionoideae—members of the basal subfamily, Caesalpinoideae, are mostly non-nodulating (Nod−). Thus, nodulation and presumably nitrogen-fixing ability are not 100% correlated even within the legume family. Nodulation may have originated multiple times in the Fabaceae: once in the only caesalpinoid that is confirmed to be nodulated, Chamaecrista genus; once in the mimosoid line; and lastly, at the base of the papilionoid line. Alternatively, there may have been a single origin of nodulation with multiple losses. In any case, more than 90% of the Papilionoideae and Mimosoideae are nodulated, whereas less than 25% of the Caesalpinoideae form nodules.

Moreover, other plant families can establish interactions with nitrogen-fixing bacteria exclusive of rhizobia. Members of eight different families, known as actinorhizal plants, are nodulated by Frankia spp. nitrogen-fixing actinomycetes. Various grasses, including such agronomically important ones as sugarcane (Saccharum officinarum), maize (Zea mays), and rice (Oryza sativa) associate with different nitrogen-fixing bacteria, among them species of Glucoacetobacter, Azospirillum, Herbaspirillum, and Azoarcus; these associations do not, however, result in nodule formation. Some of the positive responses on plant growth exhibited by these so-called "associative" nitrogen-fixing interactions are due to the production of phytohormones, but nitrogen fixation has also been demonstrated. Nevertheless, what makes rhizobia and Frankia spp. different from the associative nitrogen-fixing microbes is that most of the rhizobia or Frankia spp.-fixed nitrogen is transferred to and assimilated by the plant for the plant's growth.

The legumes and their association with Rhizobium spp. in the broad sense have always been extremely important agronomically. The use of crop rotations to enhance the productivity of nonlegume crops was vividly described by the Romans, who were probably aware of an even older tradition in Greece. Moreover, this nonpathogenic association between prokaryote and eukaryote is a fascinating phenomenon for investigation of basic biological principles.

Legumes are a Unique Family

The evidence for the evolution of the legumes (Fabaceae), the third largest family of flowering plants, is fragmentary, at least based on fossil evidence. There are no obviously identifiable nodules associated with fossils that can be accurately described as legume roots. The mostly leaf fossils date from the Cretaceous era, which has been variably dated as 65 to 145 million years ago (MYA). Thus, we do not know how long ago the first legumes started to associate with rhizobia.

A phylogenetic analysis using a chloroplast gene sequence showed that the legumes and actinorhizal plants (nodulated by Frankiaspp.) belong to the Rosid I clade, and suggested that there was a single origin for a predisposition for nodulation in this lineage. However, it is unclear as to what this predisposition entails. Does it mean that the plants have unique receptors or unusual cell walls? Do these plants produce certain types of signal molecules to entice the symbiont or to repress various types of defense molecules, thus enabling the symbiosis to occur? Do they have different phytohormones or phytohormone sensitivities? The Soltis et al. study used rbcL, and other organellar sequences have been utilized as well to study the relationships of angiosperm genera. Would nuclear gene sequences generate the same results? If there was a predisposition for nodulation, then why do the vast majority of the plants in the Rosid I clade not associate with nitrogen-fixing organisms?

If all nodules are derived from a common progenitor, how do nodules of the legumes differ from those of other plant groups? Although the ontogeny of the various actinorhizal nodules is not identical, the nodules are developmentally and anatomically more related to lateral roots than are legume nodules. Nevertheless, the legume nodule shares more traits with a lateral root than with any other plant organ. Legume and actinorhizal nodules can be indeterminate, growing by means of an apical meristem, but determinate nodules, those lacking a persistent apical meristem, are only found in legumes. Moreover, some legumes such as lupins or Sesbania rostrata develop nodules that fall into an intermediate category. Unlike the lateral root that is initiated from cell divisions in the pericycle, the legume nodule originates from cell divisions in the outer or inner cortex, depending on whether a determinate or indeterminate nodule is formed.

Flavonoids: Signals and Modulators of Nodule Development

More than 4,000 different flavonoids have been identified in vascular plants, and a particular subset of them is involved in mediating host specificity in the legumes. All flavonoids consist of two benzene rings linked through a heterocyclic pyran or pyrone ring. Specific substitutions on the ring produce flavonols, flavones, flavanones, as well as isoflavonoids, which are derived from a migration of the B ring from the 2 to the 3 position. Isoflavonoids are limited to the legume family. Daidzein and genistein, isoflavonoids produced by soybean (Glycine max), are effective inducers of Bradyrhizobium japonicum nod genes, but inhibit S. meliloti nod gene expression. S. meliloti nod genes can be induced by luteolin. This specificity enables rhizobia to distinguish their hosts from other legumes. The specific flavonoid not only induces nod gene expression, but also rhizobial chemotaxis. Nevertheless, other than the isoflavones, most flavonoids are not unique to legumes. How do soil rhizobia recognize their host and initiate the symbiosis when nonlegume plant species growing in the same area are also sources of flavonoids? Apparently, it is the next stage, once the flavonoids are perceived, where another level of specificity comes into play.

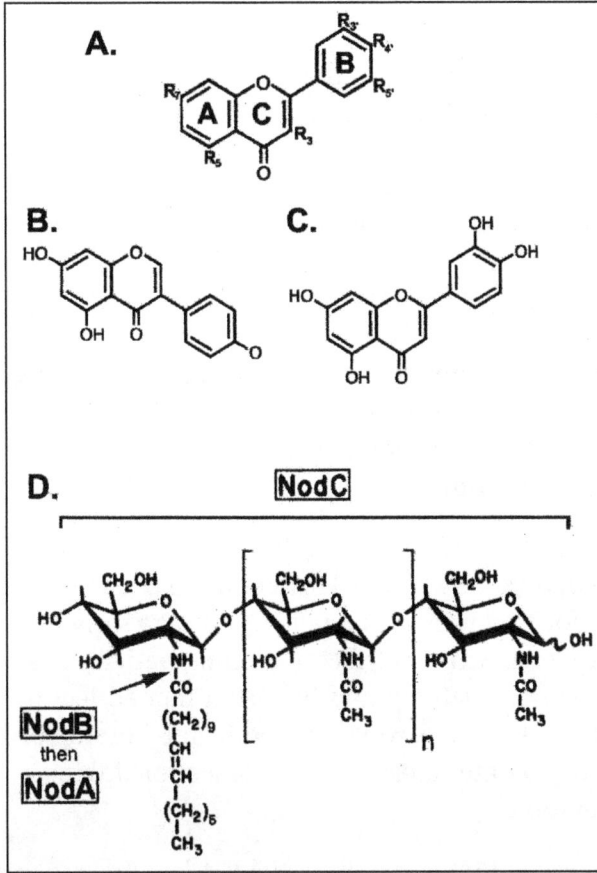

A, Generalized structure of a flavonoid. Changes on the ring or in the R groups result in flavonols, flavones, flavanones, and glycosylated flavonoids, among others. B, An isoflavonoid inducer, genistein. C, A flavone inducer, luteolin. D, Generalized structure of a Nod factor. NodC, a β-glucosaminyl transferase, links the UDP-N-acetyl glucos-amine monomers into a chitin-like backbone. NodB removes an acetyl group from the terminal residue of the chitin oligomer. Then, NodA catalyzes the transfer of a fatty acyl chain onto the resulting free amino group, using acyl-ACP from fatty acid biosynthesis.

Flavonoids are perceived as aglycones, which induce rhizobialnod genes by interacting with the gene product of nod D, a LysR-type regulator. This interaction results in a conformational change in the NodD protein such that it binds tonod box elements in the promoters of the nodgenes. The concerted expression of these genes leads to the synthesis of Nod factor molecules—lipochitooligosaccharides that usually consist of four or five N-acetylglucosamines, β-1–4 linked, with the terminal nonreducing sugar N-acylated with a fatty acid of 16 to 18 carbon residues. Nod factors can be chemically modified with acetate, sulfate, or carbamoyl groups, or can have different sugars, such as Ara, Man, Fuc, or substituted Fuc. The degree of saturation of the acyl tail can also vary. The assemblage of these substitutions result in a specific Nod factor that is recog-nized by a particular legume.

Nod Factor Responsiveness of Legumes

One of the key traits that differentiates the nodulating legumes from other plant species is their responsiveness to Nod factor. Early responses to Nod factor include ion flow across the plasma membrane and an associated depolarization of the membrane, followed by periodic oscillations in intracellular calcium, referred to as "calcium spiking"; these are followed by deformation of root hairs and initiation of cortical cell division. Root hair curling (which involves entrapment of the bacteria) and infection thread growth require the presence of the bacteria. Nodulation appears to have an absolute requirement for Nod factor because rhizobia that do not synthesize Nod factor do not nodulate, and legume mutants that are incapable of perceiving Nod factor or transducing it along a signal transduction pathway are Nod⁻. The identity of a Nod factor receptor(s) in legumes is unknown, but research is well under way. Its identity may be key to finding the "Holy Grail."

A biochemical approach has led to the characterization of high-affinity binding sites for Nod factors. One of these, NFBS2, is located in the plasma membrane and exhibits different selectivities for Nod factors in alfalfa (Medicago sativa) and bean. In Dolichos biflorus, an unusual lectin with Nod factor-binding activity has been characterized. This protein has an apyrase activity and has been named lectin nucleotide phospho hydrolase. It is not clear, however, whether Nod factor-binding proteins are unique to legumes. Furthermore, to date, there is no evidence for linking a Nod factor-binding protein to a Nod⁻ mutation.

The study of Nod⁻ plant mutants has also yielded leads for identifying the proteins involved in Nod factor perception and signal transduction. Many Nod⁻ mutants have been identified in commercially and agronomically important legumes such as pea, bean, alfalfa, sweet clover, and others, and more recently, model legumes such as Medicago truncatula and Lotus japonicus have been used for genetic studies. Transposon tagging in L. japonicus led to identification of NIN (nodule inception), which encodes a transcription factor, and the first cloned gene that is directly involved in nodule development. nin mutants showed root hair curling in response to rhizobia, but did not develop infection threads or nodules. In recent unpublished data, G.B. Kiss and colleagues positionally cloned a gene called NORK (nodule receptor kinase) from an alfalfa Nod⁻ mutant (MN1008) that shows neither Ca²⁺ spiking nor root hair deformation in response to rhizobia. Kiss and colleagues did a chromosome walk to the Nn1 locus (mutated in MN1008) using a combination of bacterial artificial chromosome clones from M. truncatula (a diploid) and markers from alfalfa (a tetraploid). They identified the mutated NORK gene in a position equivalent to the previously mapped Sym19 locus (pea) and also to the Dmi2 locus. NORK encodes a Leu-rich repeat kinase that could be a receptor, but so far it has not been shown whether or not it directly interacts with Nod factor. It is possible that NORK could interact with a Nod factor-binding protein via its Leu-rich repeats; a postulated position for this locus in nodulation signaling is shown in figure.

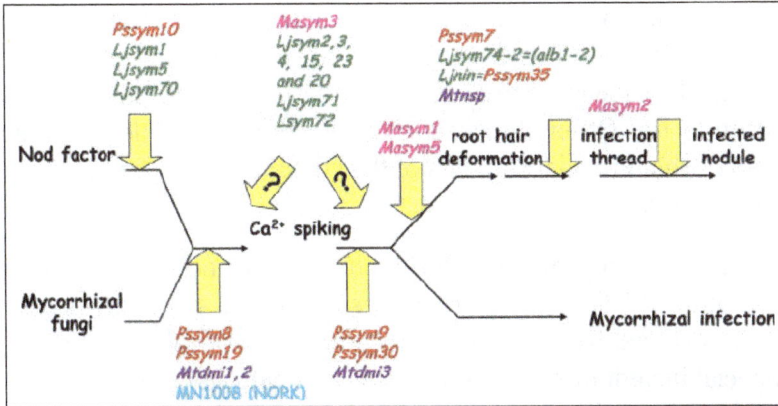

Analysis of various papilionoid legume non-nodulation mutants for their mycorrhizal phenotypes. A common pathway is observed where the two lines converge. The mutants are ordered on this pathway according to their Ca^{2+} spiking response, where known. Some of these mutants have not been tested for Ca^{2+} spiking response yet, nor have allelism tests been performed. The Masym1 mutants are mostly Myc⁻, but one mutant allele, BT62, which is leaky, is Myc⁺. The one Masym5 mutant is also Myc⁻. The mycorrhizal phenotypes of Pssym7 and Mtnsp have not been reported. Except for Ljnin, none of the genes have been cloned. The diagram is based on that of Walker et al. Green, L. japonicus; red, pea; pink, Melilotus alba; blue, M. truncatula; turquoise, alfalfa. The yellow arrows indicate the approximate stage where the mutants are blocked.

The Mycorrhizal Connection: Not Specific to Legumes, but a useful Correlation

Since the original coupling of Nod⁻ and Myc⁻ (inability to establish a mycorrhizal association) in pea and Vicia faba mutants by Duc et al., a number of papilionoid legumes that normally nodulate have been shown to be both Nod⁻ and Myc⁻. Figure illustrates the Nod⁻ Myc⁻ connection for L. japonicus (Lj), pea (Ps), alfalfa (Ms), M. truncatula and white sweetclover). A Myc⁻ mutant allele has also been described for bean. The various mutants can be described as those that are blocked very early in the signaling cascade (before Ca^{2+} spiking) and those blocked later (after Ca^{2+} spiking). Those Nod⁻ Myc⁻ legume mutants blocked before Ca^{2+} spiking are presumed to be altered in a receptor that is common to the mycorrhizal and nodulation pathways (Pssym8, Pssym19, Mtdmi1, Mtdmi2, and MN1008). A group of mutants, Masym3 and the L. japonicus sym mutants, have not been tested yet for Ca^{2+} spiking, or if they have, not all mutant alleles have been examined. Therefore, it is not known whether they are blocked before or after Ca^{2+} spiking. In addition, the results on the L. japonicus sym mutants have been generated from several different laboratories, so it is not yet known whether some of the mutations are allelic.

Pssym10, Ljsym1, Ljsym5, and Ljsym70 mutants are Nod⁻Myc⁺ and presumably blocked before Ca^{2+} spiking (to date, the Ljsym mutants have not been tested) and upstream of the NORK-type receptor. This upstream gene(s) could encode a Nod factor-binding protein and/or a nodulation specific signaling protein. Also, many legume mutants that are very likely blocked after Ca^{2+} spiking, but before infection thread formation, have been found. They include those mutated in NIN, which has been characterized as a transcription factor. Others blocked after infection thread development may have mutations in genes that encode elements of the signal transduction pathway leading to nodule morphogenesis.

The connection between mycorrhizal development and nodulation as well as the fact that the early nodulin (ENOD) genes are expressed in both symbioses suggest that nodulation may have evolved from the more ancient mycorrhizal condition. The legumes have given a special insight into the mycorrhizal association by enabling identification of genes required for this symbiosis. It is predicted that other mutations are likely to affect the mycorrhizal symbiosis, but not nodulation. However, such mutants have yet to be described. The mycorrhizal association is believed to have originated more than 400 MYA based on fossil evidence, and so it is possible that the nodulation symbiosis may have adapted some components of a much older symbiotic pathway.

Specificity of Legume Lectins

For the infection thread to form, there has to be an intimate connection between the rhizobial cell surface and the plant cell wall. Based on the strong correlation between the inoculation specificity of bacteria of the family Rhizobiaceae on their legume hosts, and the ability of host-produced lectins to bind to Rhizobium sp. cells, the lectin recognition hypothesis was formulated to explain why alfalfa and S. meliloti or soybean and B. japonicum or any other legume and its nitrogen-fixing rhizobial species are symbiotic partners. Lectins frequently follow the various cross inoculation groups established by their host legumes due to their different carbohydrate-binding specificities. Soybean lectin (SBA or SBL), a galactosamine-binding protein, differs from pea lectin (PSA or PSL), a Glc-/Man-binding protein, and they both differ from other legume lectins.

Could the lectins that appear to be characteristic of their legume hosts be involved in infection thread formation or nodulation? Work on transgenic legume plants carrying a foreign lectin gene strongly suggested that the introduced lectin enhanced rhizobial attachment, infection thread formation, and nodulation in response to heterologous rhizobial strains. However, the heterologous rhizobia must produce the compatible Nod factor for the host legume; otherwise, no nodules develop. The requirement for the compatible Nod factor suggests that the introduced lectin may be facilitating bacterial attachment, in so doing causing a localized increase in Nod factor concentration at the site of bacterial entry. However, lectins do not bind Nod factors, so the introduced lectin must be interacting with some other component(s) of the rhizobial cell surface. Different rhizobial strains have characteristic cell surfaces consisting of capsular polysaccharide, exopolysaccharide, and lipopolysaccharide. Neither Bradyrhizobium elkanii USDA31, which does not bind SBL, nor exoB mutants of B. japonicum attached to Lotus corniculatus roots or induced nodules over the non-transgenic and vector control levels. Similarly, inoculation with an exopolysaccharide-deficient mutant of R. leguminosarum bvviciae did not result in infection threads or nodules on transgenic alfalfa plants carrying the PSL gene. These data suggest that some component, which is missing or changed in the B. japonicum or R. leguminosarumbv viciae exo mutants, may be a ligand for the introduced lectin. Nevertheless, a larger question that remains is whether the legume lectins are absolutely essential for nodulation.

Legume Genes and Gene Regulation: Unique Domains

A number of early nodulin (ENOD) and nodulin (NOD) genes have been identified based on what was thought to be their exclusive expression in the nodule. However, it has now become clear that many of these genes are in fact expressed in nonsymbiotic tissues and/or during nonsymbiotic conditions. For example, ENOD40 is an early nodulin gene induced within hours of Rhizobium sp. inoculation and its expression appears to be critical for proper nodule development. However, ENOD40 transcripts are also found localized in the stele of the stem, lateral roots, and in other tissues. In addition, ENOD40 homologs have now been identified in nonlegumes, including rice, a monocotyledon outside the Rosid I clade, although so far not in Arabidopsis. Similarly, plant hemoglobins were long thought to be nodule-specific proteins, but homologs have now been found in rice and Arabidopsis, among others.

There are some nodulin genes that appear to be novel, such as some of the peribacteroid membrane proteins, which may have originated due to gene duplications and/or recombination. However, it seems that many of the genes involved in nodule development and nitrogen fixation were recruited from their original task in plant growth and development to function in the nodule. Understanding how these genes are regulated may contribute to our understanding of what makes legumes unique. Recent data indicate that some of the regulatory genes have domains that may be found exclusively in legumes, such as the Hy5 homolog, LjBZF and a DNA-binding protein, VsENBP1. Are there other genes with legume-specific regulatory domains and are these regions critical for nodulation?

How did bacteria acquire the ability to establish a symbiosis with legumes? In the absence of a bacterial fossil record, it is difficult to date speciation within bacteria. However, analysis of evolutionary changes in highly conserved genes can be used as a "molecular clock." Such studies suggest that the fast-growing rhizobia (e.g. Rhizobium sp. and Sinorhizobium) diverged around 200 to 300 MYA, whereas divergence between fast-growing rhizobia and slow-growing bradyrhizobia occurred around 500 MYA. These times are earlier than the split between monocots and dicots (156–171 MYA) and the separation of brassicas and legumes (125–136 MYA). Therefore, rhizobia appear to have diverged well before the existence of legumes and probably before the appearance of angiosperms. Therefore, nodulation capacity is thought to have been acquired after bacterial divergence and horizontally spread among different genera.

This concept is strongly supported by the recent finding that Burkholderia strain STM678 can nodulate legumes. This genus is in a completely different subdivision (β) of the proteobacteria from the rhizobia (α-subdivision), and so these bacteria are essentially unrelated. Nevertheless, the nodulation genes are clearly similar to those from rhizobia. The question as to where "rhizobia" evolved from can be restated as: "What are the unique elements that enable rhizobia to establish a symbiosis" or, perhaps more specifically, "where did the nodulation genes come from"?

Nodulation Genes are Unique Qualities of Rhizobia

The ability to fix atmospheric N_2 is very widespread among bacteria and Archaea, although interestingly, this capacity is restricted to prokaryotes. Therefore, there are many different diazotrophs that, if equipped with the ability to invade plants, could theoretically evolve to establish a nitrogen-fixing symbiosis. One change, however, would be to uncouple the regulation of nitrogen fixation in planta from the microbial requirement for fixed nitrogen, something rhizobia have done very efficiently.

There are many bacteria that grow endophytically within plants, but what distinguishes the rhizobia is their ability to make "Nod factors," molecules required to program the specialized infection process and nodule morphogenesis. The biosynthesis of Nod factors has been thoroughly reviewed. Although Nod factors can carry many substituents, which are important for nodulating specific legumes, their basic structure requires the action of only three gene products, NodA, NodB, and NodC. NodC is an N-acetyl-glucosaminyl transferase that produces the chitin backbone from UDP-N-acetyl glucosamine. NodB removes an acetyl group from the terminal residue of the chitin oligomer, and NodA catalyzes the transfer of a fatty acyl chain onto the resulting free amino group, using acyl-ACP from fatty acid biosynthesis.

The origin of the nodA, nodB, and nodCgenes therefore may be crucial. It is likely that they came from outside the Rhizobiaceae because, like most of the nodulation and nitrogen fixation genes, they have a G + C content that is significantly lower than the average G + C content of rhizobia; they also have a different codon usage from most chromosomal genes. NodC is one of a large class of bacterial β-glucosyl transferases, many of which can incorporate N-acetyl glucosamine into cell wall polysaccharides. For example, Streptococcus pyogenes produces a polymer of alternating β-1,4-linked GlcNAc and GlcUA. Furthermore, the peptidoglycan of many bacteria is composed of a backbone of alternating β-1,4-linked GlcNAc and N-acetyl muramic acid (which is the lactic acid ether of GlcNAc). It is possible that a NodC-like protein could have evolved from such a bacterial enzyme. There are several NodB-like proteins in databases, and it is easy to imagine how a simple glucosamine-deacetylase like NodB could have been recruited.

The potential origin of NodA is an enigma and its function is unusual because it adds a fatty acyl chain to a preformed polysaccharide. Almost all bacterial fatty acylated polysaccharides studied are produced by incorporating acylated sugars during elongation of the polysaccharide. NodA-like proteins are special because thus far they have been found only in rhizobia and no related proteins are detected in database searches. Perhaps these nodulation genes came from some bacterial source that has yet to be sequenced. The unusual characteristics of NodA may enable us, in the future, to get an insight into what that source may have been.

An alternative view is that the key nodulation genes may have been acquired from fungi. Most fungi make chitin as part of their cell wall and therefore have chitin synthases, which are similar to NodC. Some fungi contain endosymbiotic bacteria. More significantly, one of the endomycorrhizal fungi, which can infect plant roots using a pathway that seems to share steps in common with nodulation, was found to contain a Burkholderia strain that harbored nitrogen fixation genes. This, taken together with the finding that a related Burkholderia strain can nodulate, may be a significant coincidence. However, Burkholderia spp. typically have a G + C content similar to rhizobia and so are unlikely to be the source of the low G + C symbiosis genes found in rhizobia.

Role of Rhizobial Genomic Sequences

The complete sequences of S. meliloti and Mesorhizobium loti have recently been completed and provide a wealth of data. Both genomes are large (6.7 and 7.6 Mb, respectively) and there is clustering of many genes known to be required for the symbiosis. In M. loti, many of the symbiosis genes are located on a chromosomal symbiosis island of 611 kb, whereas in S. meliloti, most of the symbiosis genes are located on either of two large plasmids, pSymA (1.35 Mb) or pSymB (1.7 Mbp). The location of symbiosis genes on "islands" or plasmids reinforces the idea that these regions have the potential to be horizontally transferred. Although the pSym plasmids of S. meliloti are not transmissible, they are clearly related to other highly transmissible plasmids. Earlier work on the symbiosis island of a M. loti strain demonstrated that this was an exceptionally efficient mechanism of transferring nodulation capacity to Nod– bacteria in field experiments. The mechanism of excision and integration of the symbiosis island out of and into the chromosome has been established to occur via integration into a Phe-tRNA.

It is surprising that 35% of M. loti genes have no orthologs in S. meliloti, and this diversity is further exemplified by the finding that over 50% of the genes on the 536-kb symbiosis plasmid of NGR234, a strain very closely related to S. meliloti, have no orthologs in S. meliloti. In fact, the most different region in the comparison of the predicted gene products of M. loti with S. meliloti corresponded to the symbiosis island. This suggests that although some very highly conserved nodulation and nitrogen fixation genes are required for symbiotic nitrogen fixation, many different genes are specifically required to optimize interactions with different legume hosts. Therefore, it is difficult to generalize, although it is evident that rhizobia have numerous solute transporters and are rich in catabolic genes, presumably enabling them to compete successfully in the rhizosphere and in soil.

The uneven distribution of insertion elements, intergenic mosaic elements, percent G $^+$C, and altered codon usage on pSymA shed light on the evolution of S. meliloti. Thus, a typical aerobic heterotrophic bacterium may have first greatly extended its metabolic potential by acquisition of pSymB. The subsequent gain of pSymA conferred the ability to infect plants, form nodules, successfully colonize the low oxygen environment of the nodule, and thereafter fix nitrogen. However, both S. meliloti and M. loti seem to have acquired highly evolved symbiotic gene packages, and so we are still left with the conundrum about how the process originally started.

Cell Division and I.E. making a Nodule

Most of the research on rhizobia-legume symbioses has focused on papilionoid legumes and their symbionts, many of which have been selected for agronomic performance. Thus, several discoveries relevant to papilionoid legumes may not apply to all symbiotic interactions, particularly for understanding the evolution of nodulation. For example, extended infection threads are required for pea and alfalfa nodulation, but there are examples where infection threads are almost nonexistent and bacteria spread interstitially as in peanut. In some tropical legumes and also in Parasponia sp., rhizobia are not released into membrane-bound symbiosomes; rather, they fix nitrogen within specialized fixation threads. Is fixation thread development and nodule morphogenesis a prerequisite for this nitrogen-fixing symbiosis or could accumulations of bacteria between cells, such as what occurs in associative nitrogen-fixing interactions, have provided fixed nitrogen in primitive, evolving symbioses? There are reports of nitrogen-fixing bacteria, based on acetylene reduction assays, in more basal legumes, but these studies remain preliminary.

Do any of the legume or rhizobial mutants characterized so far shed light on which genes can be dispensed with, yet allow the symbiosis to proceed in what might be considered akin to the primitive condition or ground state? Several of the mutations affecting host-specific modifications of Nod factors delay nodulation, but in many cases, the process continues normally. However, an interesting phenotype was described for a mutant of R. leguminosarum bv viciae lacking all of the host-specific nodulation genes but retaining the nod ABC genes and their regulator. On vetch (Vicia sativa), many hundreds of root hairs were heavily infected, but infection threads and nodules were not formed. If similar levels of infection of root hairs on an evolutionarily more basal legume were to occur, and nitrogen fixation could take place within these infected cells, then we could postulate that there might be the potential to provide significant levels of nitrogen to the plant. The ability of many Nod$^-$ legumes to accumulate high levels of nitrogen could argue positively for some sort of non-nodular association with rhizobia. Alternatively, these plants may be efficient nitrogen scavengers.

The potential to induce cell division and create a nodule greatly enhances the efficiency of the symbiosis. It may be significant that minimalist Nod factor structures can induce early signaling events, whereas more highly substituted Nod factors are required to initiate cell division, nodule primordia, and infection thread structures. This has led to the idea that there may be different levels of recognition of Nod factors. Some plant genes involved in processes related to cell division, such as cell cycle control and nuclear endoreduplication, have been identified, and these may act relatively late in relation to the developmental scheme briefly sketched in figure. Pingret et al. suggested a role for a G protein-mediated signaling pathway for induction of legume early nodulin genes, based on inhibitor studies and the induction of gene expression by mastoparan, a G protein agonist. However, mastoparan did not induce calcium spiking in root hairs, and taken at face value, this would imply that a role for G protein-mediated signaling could be downstream of calcium spiking.

Actinomycetes

Actinomycetes are a class of bacteria. They are highly beneficial. Numerous antibiotics are created from the bacteria. The bacteria also form a symbiotic relationship with many types of plants. Together with the plant, the Actinomycetesbacteria successfully fix nitrogen in the soil, which the plants use to live, which enables the bacteria to continue to thrive.

Actinomycetes also help to break down organic material to enrich soil. Bark, plant fiber, and the exoskeletons of numerous insects are readily consumed by the bacteria, which makes them optimal natural decomposers in the garden.

Actinomycetes bacteria form long filaments. The filaments stretch through the soil. The Actinomycetes bacteria is almost fungi-like and work together to control harmful or unwanted soil bacteria.

When scientists discovered that Actinomycetes bacteria controlled harmful bacteria, they promptly designed an antibiotic from the bacteria on October 19, 1943. The bacteria-laden antibiotic has the ability to control many penicillin-resistant bacteria. The antibiotic was named Streptomycin and was the first drug ever used to cure tuberculosis. Further antibiotics were developed from the bacteria, such as Erythromycin, Neomycin, Tetracycline, and Cefoxitin.

The use of beneficial bacteria in gardening is considered an organic approach to gardening, and its application is becoming more popular as gardeners race to become more organic and sustainable.

An easy way to add some of these beneficial bacteria to your garden is to look for some bokashi, a type of compost. Bokashi typically contains tons of actinomycetes that boost oxygen levels in soil systems and even in hydroponic systems, which typically have a low microbial diversity.

Characteristics of Actinomycetes

The Actinomycetes or Streptomycetes or Actinomycetales as they are called are a group or Gram-positive bacteria which form branched filamentous hyphae having resemblance with fungal hyphae. But their hyphal diameter is approximately 1μm, whereas in fungi it is 5 to 10 μm.

These organisms reproduce by asexual spores which are termed conidia when they are naked or sporangiospores when enclosed in a sporangium. Although these spores are not heat-resistant, they are resistant to desiccation and aid survival of the species during periods of drought.

These filamentous bacteria are mainly harmless soil organisms, although a few are pathogenic for humans (Streptomyces somaliensis causes actinomycetoma of human), other animals (Actinomyces bovis causes lumpy-jaw disease of cattle), or plants (Streptomyces scabies causes common scab in potatoes and sugar beets).

In soil they are saprophytic and chemoorganotrophic, and they have the important function of degrading plant or animal resides.

Again some are best known for their ability to produce a wide range of antibiotics useful in treating human diseases. These organisms excrete extracellular enzymes which are decomposers of dead organic material. These enzymes lyse bacteria and thereby keep the bacterial population in check and thus help to maintain the microbial equilibrium of the soil.

The Actinomycetes superficially resemble fungi for having subterranean and aerial hyphae and chains of spores. But their hyphal diameter, cytology and chemical composition of cell walls are quite decidedly bacterial in pattern.

Economic Importance of Actinomycetes

The Actinomycetes, forming soil micro-flora have gained the greatest importance in recent years as producers of therapeutic substances.

Many of the Actinomycetes have the ability to synthesize metabolites which hinder the growth of bacteria; these are called antibiotics, and, although harmful to bacteria are more or less harmless when introduced into the human or animal body. Antibiotics have in modern times great therapeutical and industrial value.

The past decade has seen considerable interest in the Actinomycetes as producers of antibiotic substances. The successful use in chemotherapy of streptomycin, chloromphenicol (Chloromycetin is the trade name of this substance), aureomycin and terramycin all metabolites of the Actinomycetes, has stimulated the search for new Actinomycetes and new antibiotics among the Actinomycetes. The genus Streptomyces is the largest and the most important one, antibiotically speaking.

Distribution and Mode of Nutrition of Actinomycetes

The Actinomycetes are essentially mesophilic and aerobic in their requirements for growth and thus resemble both bacteiia and fungi. They along with other micro-organisms, form the soil microflora and produce powerful enzymes by means of which they are able to decompose organic matter.

The majority of these are soil organisms and are associated with rotting material. The characteristic odour of soil after it is ploughed or wetted by rain is largely due to the presence of the Actinomycetes.

Some are pathogens. The Actinomycetes grow slowly and on artificial media produce hard and chalky colonies which smell decaying leaves 01 musty earth. They are par-ticularly abundant in forest soil because of the abundance of organic matter. They occur mainly in soils of neutral pH, although some prefer acidic or alkaline soil. The Actinomycetes can grow in soils having less water content than that needed for most others bacteria.

The Actinomycetes are capable of utilizing a large number of carbohydrates as energy sources when the carbohydrates are present in the media as sole sources of meta- boliz-able carbon.

Most of the Actinomycetes are quire proteolytic and attack proteins and polypeptides, and are also able to utilize nitrates and ammonia as sources of nitrogen. Nearly all syn-thesize vitamin B12 when grown on media containing cobalt salts, and many are able to synthesize rather complex organic molecules which have antibiotic properties. The mechanism of synthesis of these substances is not understood.

Somatic Structures of Actinomycetes

Most of the Actinomycetes are mycelioid. They begin their development as unicellular organisms but grow into branched filaments or hyphae which grow profusely by pro-ducing further branches constituting the mycelium. The width of the hyphae is usually 1 μm. The delicate mycelia often grow in all directions from a central point and produce an appearance that has been compared with the rays of sun or of a star.

Therefore, the Actinomycetes are also called 'ray fungi'. They often produce compli-cated designs and resemble some of the drawings in modern art exhibitions. They are Gram-positive. The protoplasm of the young hyphae appears to be undifferentiated, but the older parts of the mycelium show definite granules, vacuoles and nuclei.

Many Actinomycetes at first produce a very delicate, widely branched, mycelium that may embed itself into the soil, or, if grown in culture, into the solid medium. This kind of mycelium is therefore called the 'substratum or primary mycelium'.

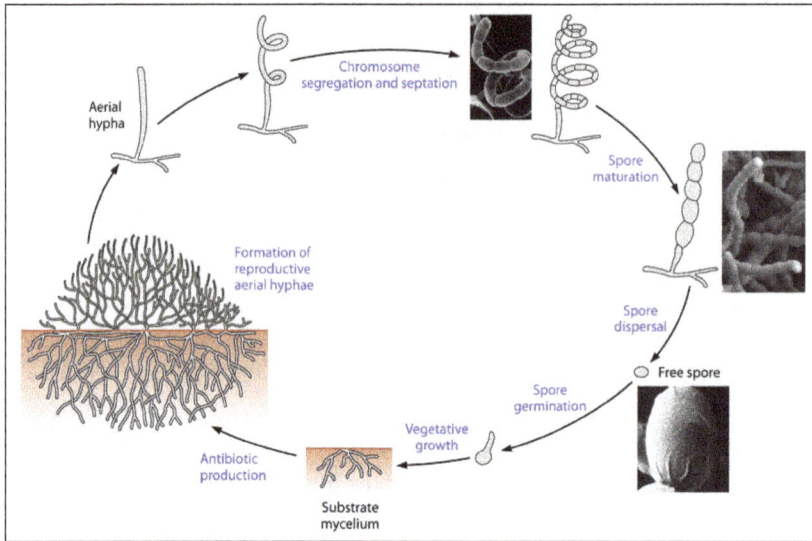

Life cycle of an actinomycete

After a period of growth, hyphae of a different kind develop, which raise themselves up from the substratum mycelium and grow into the air. These ate called aerial hyphae, and the corresponding mycelium is the aerial or secondary mycelium. The aerial mycelium may be white yellow, violet, red, blue, green, or grey and many form pigments that are excreted into the medium.

The aerial mycellium is usually slightly wider than the substratum mycelium. The aerial hyphae possess an extra ceil wall layer (sheath). The hyphal tip undergoes septation within this sheath to form a chain of conidia. Conidial cell contains a plump, deeply staining, oval or rod-shaped nuclear body.

Reproduction in Actinomycetes

Most species reproduce by conidia which are developed in chains from the aerial hyphae. The chains may be straight, flexuous (wavy) or coiled to various degrees. The conidia bearing filaments are often spirally twisted. Sometimes the whole length of the aerial hypha, sometimes only its upper part is transformed into conidia.

Each conidium has a roundish nucleus and is surrounded by a firm outer wall. The conidial wall may be smooth, warty, spiny, or hairy.

The conidia can persist in the dry state for many years. Even the vegetative forms of the Actinomycetes are quite hardy and are able to adapt themselves to the changing soil conditions.

The conidia appear as a fine powdery coat on the surface of cultures. When the conidia have been scattered on the ground and conditions are favourable they germinate producing one to three or even occasionally four little germ tubes which give rise to mycelioid condition.

The primary mycelium in some species commonly breaks up into small fragments called arthrospores, which often look like bacterial cells and which might easily be mistaken for the latter.

Archaea in Dry Soil Environments

Archaea belong to the least well known major group of soil inhabiting microbes as the concept of the very existence of the archaea was introduced only in 1977 and the domain of Archaea established in 1990.

Frequency of Archaea in Soils

The number of prokaryotes on earth has been estimated to be 1028–1030. The total amount of carbon contained by the prokaryotes is 60–100% of that contained by plants and they represent the largest pools of N and P in living organisms. Soils contain roughly 5% of all prokaryotic life forms. Although soils are lower in microbial cell numbers than some other habitats, such as oceanic sediments and subsurfaces, soils have extraordinarily diverse microbial populations. One 10g soil sample can contain 103–107 different microbial species or genotypes. Of the prokaryotes living in moderate aerobic soils approximately 0.5–3.8% are archaea. Most commonly the estimates vary between 1 and 2%. Some studies, however, show only low ratios or no archaea even in undisturbed soils. In a pyrosequencing study Roesch et al. showed only 0.009% archaea of all prokaryotes in a forest soil, but it is unclear, if more than one 1g sample was taken from the notoriously heterogeneous soil archaeal community Liles et al. did not find any evidence of archaea from undisturbed soil in Wisconsin, but they screened for 16S rRNA genes through bacterial artificial chromosome (BAC) clones constructed in Escherichia coli. As archaeal fragments are often quickly expunged from E. coli cells, the result is expected, but does not report the situation in the soil. The highest estimates of percentages of soil archaea (12–38% of 16S rRNA gene pool) were reported by Kemnitz et al., but the primers used in the study may have also amplified some types of bacterial 16S rRNA gene sequences and the standardisation of the qPCR is not reported in enough detail to validate the archaeal bacterial rates. All, archaeal 16S rRNA gene sequences have been ubiquitously retrieved from upper soil layers from all continents whenever they have been thoroughly looked for.

Crenarchaeota in Soils

Most Crenarchaeota found thus far from moderate soil environments have belonged to four distinct clades; Group I.1a, group I.1b, I.1c and I.3, but a few crenarchaeotal sequences have also fallen outside these clades. Group I.1b Crenarchaeota appear to be the most widely spread and common soil archaea. They dominate or are the only type of archaea found in many different types of soils. Additionally dwarf Crenarchaeota of group I.1b have been found from semiarid shrub land soil. Group I.1b Crenarchaeota are also frequently found from fresh water. Group I.1c Crenarchaeota are common i highly organic and acidic soils. They are thus far the only crenarchaeotal type discovered in boreal forest humus, but have been reported also from other soils rich in organic matter or from soils with low pH. According to current knowledge Groups I.1a and I.3 Crenarchaeota appear to be more rare in the soils than the groups I.1b and I.1c. Some Crenarchaeota clustering outside group I.1a with sequences previously found in wetlands, hot springs, uranium mines or coal seams have also been found from 0.5 to 1m depth of temperate deciduous forest soil and B-horizon of loamy soil in Tennessee.

Euryarchaeota in Soils

The diversity and frequency of Euryarchaeota in moderate dry oxic soils appears to be much lower than that of Crenarchaeota. Methane producing microbial populations were successfully grown from five types of oxic soils (forest, cultivated field, termite soil, savannah, desert) by Peters and Conrad. After an initial lag period there was potential for methane production in anaerobic conditions in all the tested soils. Methane production was also observed in beech forest soil. Indication of methanogenic Euryarchaeota was also found in compacted inter row soil of a potato field when phospholipid etherlipid derived isoprenoids were analysed by Gattinger et al. Some methanogenic Euryarchaeotal 16S rRNA gene sequences have been found from Swedish barley field agricultural sandy soil.

Euryarchaeotal 16S rRNA gene sequences associating with Thermoplasmatales have been found in several types of soils. The first report is from temperate mixed Norway spruce-beech forest topsoil in Switzerland. Thermoplasmatales have also been found in temperate deciduous forest soil in Germany and from different Australian topsoils. A gene bank submission of 16S rRNA genes similar to Halobacterium salinarum retrieved from boreal forest soil has been made by Jurgens et al. Members of Korarchaeota and Nanoarchaeota have not been detected in soil. These are small groups and may be rarely or not at all present or they do not amplify with the commonly used PCR primers.

Soil Depth and pH

A decrease of both Euryarchaeota and Crenarchaeota has been observed with increasing soil depth in mixed Norway spruce-beech forest. A similar trend was observed by Kemnitz et al. who estimated that the overall number of archaeal 16S rRNA genes per gram of forest soil was higher in the top 10 cm than in the lower 10–20 cm soil profile, the maximum numbers being 3.8 910816S rRNA genes per gram of dry soil. However, the percentage of archaea of the total soil prokaryotes increased, up to 38%, in the deeper soil layers as the number of bacteria decreased. Some archaea became more rare and some more abundant with increasing soil depth showing adaptation to different depths and conditions in a study by Bundt et al. Hanselet al. found that archaeal populations were dissimilar in different soil layers, but concluded that these variations appeared to be a result of soil pH rather than depth. In contrast to these studies Poplawski et al. did not find depth to have any effect on the archaeal community structure.

Group I.1c Crenarchaeota appear to prefer acidic soils. In fact I.1c Crenarchaeota have been reported from soils with pH above 5.1 only once, and this was a study where originally acidic boreal forest soil was amended with wood ash. Group I.1b and I.1c Crenarchaeota have been reported from the same soils, but usually I.1b Crenarchaeota are detected from soils with pH higher than 5.

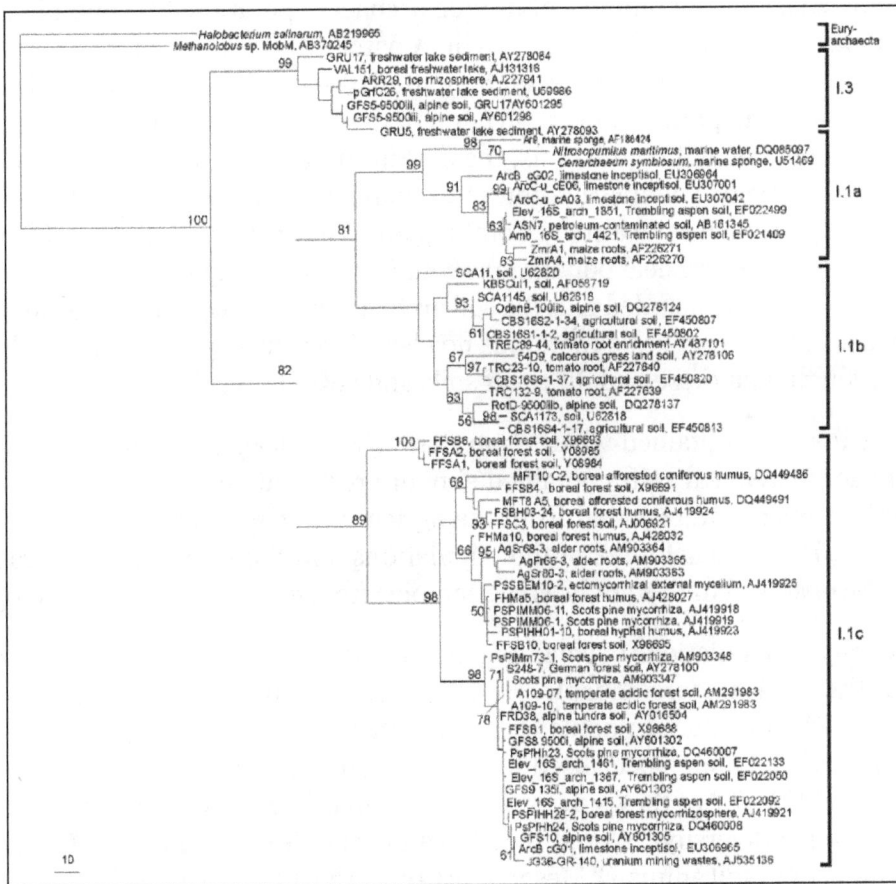

Phylogenetic cladogram on 16S rRNA gene sequences of soil, rhizosphere and mycorrhizosphere Crenarchaeota. Archaeal 16S rRNA gene sequences were aligned using the Clustal alignment tool and the alignments were manually edited with the BioEdit alignment editor (version 7.0.5ÓTom Hall). Maximum parsimony analysis was performed with PAUP 4.0b (Sinauer Associates, Sunderland, MA, USA) with heuristic search parameters of 10repeats of step-wise addition of random sequence with a tree-bisection-reconnection (TBR) branch-swapping algorithm. A maximum of 10,000 trees was set for the analysis. Bootstrap support values were calculated on 1,000 boot-strap replicates. Values for nodes with over 50% support are indicated.

Heterogeneity of Archaeal Populations in Soils

Although there are some larger trends as described above, a small scale heterogeneity and patchy distribution of soil archaeal populations is evident from many studies. Nicol et al. showed that regardless of sampling distance 0.1 and 1g samples of soils had dis-similar archaeal communities, whereas the 10g soil samples were similar even if taken several meters away from each other. Archaeal communities of forest, shrub land and prairie sites were shown to be consistently less even than the bacterial communities in the same soils and the number of unique archaeal operational taxonomic units as similar or exceeded those of bacterial origin. A patchy distribution of archaea was also found by Sliwinski and Goodman when they studied six different types of soils in Wis-consin. Although the prairie, forest, turf and agricultural soils had characteristic and dissimilar crenarchaeotal populations, the distribution of the archaea within the hab-itats was very uneven. The distribution of Crenarchaeota in boreal forest humus was found to be highly stochastic also by Bomberg and Timonen. In a study by Oline et al. the differences in the archaeal populations could be observed only in the smallest scale and the authors thus concluded that the soil archaea exist in highly localized and clonal populations. In the larger scale there was no clear difference between sites at different altitudes even though these had different soils and vegetation.

Similar results were obtained when Australian agricultural, grassland and woodland sites were compared. All soils contained similar cren- and euryarchaeotal sequences, only more sequences were found from the agricultural site maybe due to the higher moisture or pH. The patchiness of the populations and the small scale heterogeneity could not be observed in this study as the samples from each site were pooled together.

Heterogeneity in populations of small organisms such as archaea in an environment as diverse as the soil is understandable. However, if soil archaeal communities are really more unique than bacterial populations it is interesting to think of reasons leading to this. Are archaea more specialised to specific habitats? Are they slower to adapt to changing conditions? Are they slower but more tenuous in lifestyle and that would al-low more unique diversity? Would they have more unique habitats in rhizospheres, with soil fungi or in soil animals? Mesophilic Crenarchaeota have already been detected in the midgut and methanogenic Euryarchaeota in hindgut of humivorous larvae of beetles. Some of the smaller soil animals and root areas as well as fungal hyphae are probably included in soil samples in a random manner and this may explain some of the discrepancies in the different studies. All in all environmental micro-organisms in general occur in non-random patterns. So the apparently stochastic and random distri-bution of soil archaea will probably be highly logical as we gain more information about the specific soil habitats and lifestyles of these organisms.

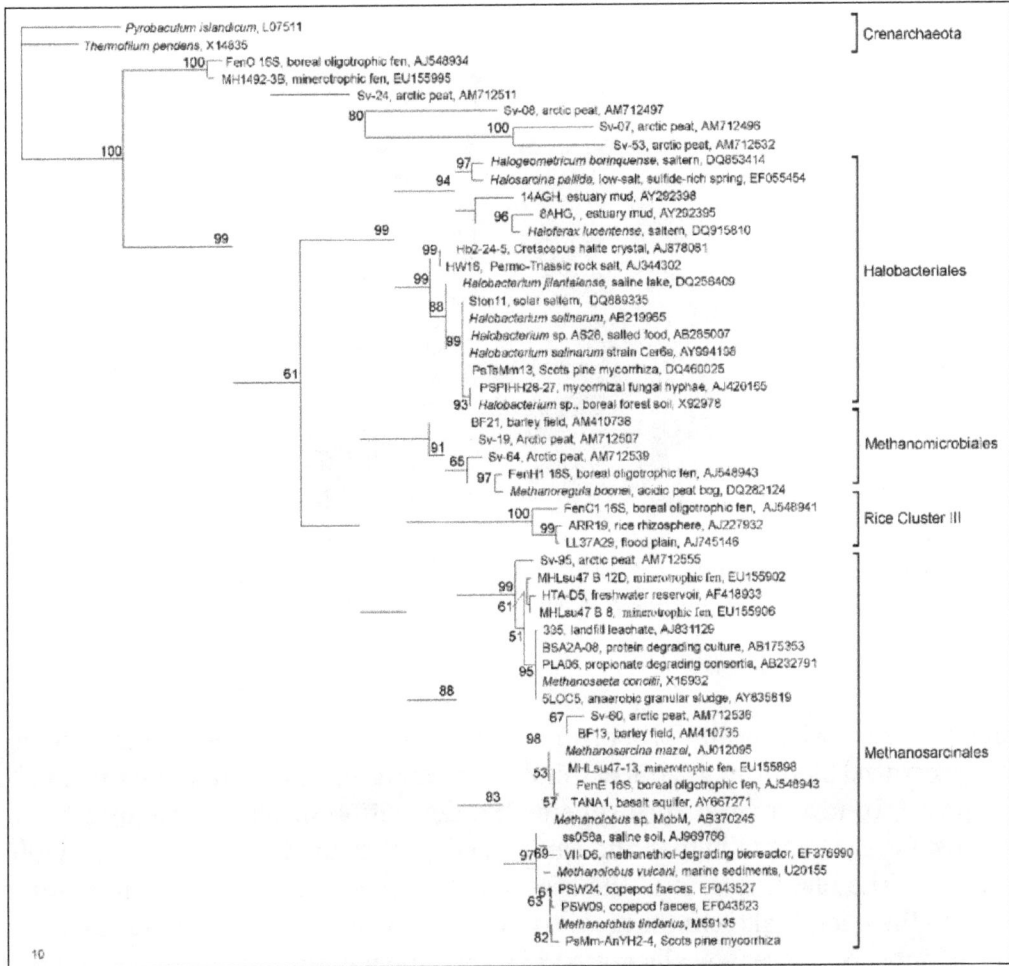

Phylogenetic cladogram on 16S rRNA gene sequences of soil, rhizosphere and mycorrhizosphere Euryarchaeota. Alignments, maximum parsimony analysis and calculation of bootstrap values were carried out. Values for nodes with over 50% support are indicated

Euryarchaeotal 16S rRNA gene sequences detected in moderate dry oxic soils:

- Methanosarcinaceae, RC1
  - Barley field soil Uppsala,
- Marine group II
  - Agricultural sandy soil South of Berlin,
- Thermoplasmatales
  - Mixed Norway spruce-beech forest top soil Unterehrendingen,
  - Temperate deciduous forest soil Marburg,
  - Topsoils New South Wales.

Scots pine-Suillus bovinus ectomycorrhizosphere grown on flat layer of forest humus. The different parts of the mycorrhizosphere; immediate rhizosphere and hyphosphere as well as soil outside the sphere of the plant-fungal influence are clearly visible in a flat microcosm.

Rhizosphere Effect

Plants deposit a significant proportion of their photosynthates to their roots. In non-mycorrhizal plants up to 20% of this carbon can be deposited into the surrounding rhizosphere. In mycorrhizal rhizospheres the deposition of plant carbon to the mycorrhizosphere is usually higher and can exceed 50% of the total plant photosynthate production. The plant cover and the associated mycorrhizal fungi thus have a major impact on the microbial communities as well as on the activity of the top soils. As 80% of land plant species are typically mycorrhizal and most natural moderate dry oxic soils are dominated by mycorrhizal plant roots and fungi it is important to consider their influence on the soil archaea.

Archaea have been reported in a host of rhizosphere studies where the mycorrhizal status of the plants has not been indicated. First root associated archaea to be reported were group I.1b Crenarchaeota from tomato roots grown in agricultural soil. Later these Crenarchaeota have been aerobically enriched in laboratory cultures, but no pure cultures are available as yet. The substrates supporting the growth of these Crenarchaeota are unknown, but they could be exudates of the tomato roots, the symbiotic mycorrhizal fungi or the associated bacteria. Maize was the next plant to have its roots studied for the presence of archaea. Crenarchaeota of the group I.1a and Euryarchaeota of Marine groupII were reported from maize rhizospheres grown in agricultural soil. Presence of archaea was indicated in potato rhizospheres by detection of phospholipid etherlipid derived cyclic isoprenoids in root containing soil of a potato field. The first root associated I.1c group crenarchaeotal 16S rRNA gene sequences were reported from non-mycorrhizal roots of Scotspine, Norway spruce, Silver birch and common elder grown in boreal forest humus. GroupI.1a and I.1c Crenarchaeota have also been reported from aspen root associated soil, but it is not quite clear, if the sequences were retrieved from soil outside the immediate rhizosphere, ectomycorrhizas or hyphosphere.

Archaeal populations of rhizospheres and soil outside them were significantly different when 76 plant samples including mosses, club mosses, ferns, conifers and seed plants and their respective soil samples were collected from their native environments. However, the different plant roots appeared surprisingly similar and only I.1b Crenarchaeota were detected in this study. Similar results were gained by Nicol et al., when they examined the rhizosphere soil of plants colonising a gradient of maturing soil in front of are ceding glacier.

In tall fescue rhizosphere soil the archaeal cell counts per gram soil were calculated to be approximately 109 and an order of magnitude lower in the surrounding soil. In this study, fungal shoot endobiotic colonisation had a decreasing effect on archaeal colonisation density of the rhizosphere soil only in the youngest seedlings and only in one of the soil types tested, but archaeal numbers consistently increased with increasing rhizosphere age. Archaeal frequencies were ten fold higher on senescent roots than on young roots also in tomato plants. The rhizosphere archaea are probably rather tightly attached to their root habitat as in alysimeter study archaea were not an abundant group in soil water and no difference was detected in soil water archaeal populations when different plants or no plants were grown in the experimental field. Nevertheless, archaea are not the dominating prokaryotes in the rhizospheres. The ratio of archaeal to bacterial 16S rDNA was 0.16% in the rhizospheres of a sandy ecosystem. This ratio is lower than in soils, probably due to the vigorous proliferation of bacteria in plant rhizospheres. Similarly Crenarchaeota were not detected from plant rhizosphere of the Proteaceae family and were only present in the non-rhizosphere soil.

Crenarchaeota in Mycorrhizospheres

The first study on verified mycorrhizal fungal effects on archaea was carried out with AM-mycorrhizal and non-mycorrhizal roots of salt marsh grass Spartinapatens. In a microscopy study with fluorescent probes archaea were associated with both mycorrhizal and non-mycorrhizal rhizospheres. However, the numbers of archaea were too low to assess, whether mycorrhizal colonisation influenced the archaeal density. The next report came from boreal forest soils where Scots pine ectomycorrhiza fine roots were shown to harbour I.1c Crenarchaeota in contrast to non-mycorrhizal fine roots. The result was confirmed a few years later when only one crenarchaeotal sequence was retrieved from non-mycorrhizal Scots pine fine roots in contrast to many different types of I.1c Crenarchaeota from roots colonised with three different species of ectomycorrhizal fungi. Similarly colonisation by anectomycorrhizal fungus Paxillus involutus increased the diversity and frequency of archaeal 16S rRNA gene sequences detected both in Norway spruce and Silver birch fine roots. In the latter study all crenarchaeotal sequences belonged to I.1c Crenarchaeota but those detected from boreal forest tree ectomycorrhizas clustered separately from the crenarchaeotal sequences retrieved from fine roots. The archaeal OTU frequencies were consistently higher in ectomycorrhizas than in fine roots or in soil outside the mycorrhizosphere. These results indicate that Crenarchaeota form a steady microbial community in boreal forest tree ectomycorrhizas. The mycorrhizal effect could be mainly due to the ectomycorrhizal fungi themselves or the effect they have on the plant exudates or on the bacteria colonising the rhizosphere. Timonen et al. and Timonen and Hurek have shown that the bacterial communities of non-mycorrhizal tree roots are different from those in ectomycorrhizalones. Archaea appear to be more frequent in the habitats where currently culturable bacteria are more in frequent. Nevertheless, archaea and bacteria are known to form syntrophic consortia with each other.

Ectomycorrhizal fungal external mycelia emanating from the mycorrhizas can reach tens of centimeters out into the soils surrounding the plant rhizospheres. The combined length of mycorrhizal fungal myceliumin organic dry soils is very high, up to 600 mg^{-1} soil. These external mycelia of several ectomycorrhizal fungi have also been shown to harbour a steady population of I.1c Crenarchaeota. In fact, I.1c Crenarchaeota have thus far been detected only from ectomycorrhizal roots or soils colonised by typically mycorrhizal plant species the only unclear publication being Hansel et al. where no description of vegetation type is given. In the succession studies from glacier for land soils I.1c Crenarchaeota only appeared when mycorrhizal plant species were already growing in the soil. They were not present in young soils inhabited only with non-mycorrhizal pioneering plants. In contrast I.1b Crenarchaeota were present in the vicinity of both mycorrhizal and non-mycorrhizal plants and regardless of the soil age. Thus it is possible that I.1c Crenarchaeota are directly or indirectly mycotrophic. Then again, some of the I.1c Crenarchaeota may respond to soil organic matter or specific decomposition products or soil pH even if part of the I.1c Crenarchaeota were mycotrophic. As the mycorrhizal fungal associated archaeal reports by Bomberg and Timonen are the first studies showing archaeal fungal associations it is difficult to know how common these associations are in soils. Nevertheless, ectomycorrhizal fungi are the dominating microbes in boreal forest soils and colonise the roots of most woody plants. Thus a large fungal biomass and a great number of different kinds of specific niches can be found particularly in organic forest soils where the amount of fungal mycelium is high. A lot of the microbial activity can also take place during winter as both fungi and archaea are known for capacity for subnival activity and considerable part of the plant litter is degraded during winter time.

Mycorrhiza derived I.1c Crenarchaeota were grown in both aerobic and anaerobic cultivations and both types of archaea fell in the same phylogenetic clusters according to their 16S rRNA gene sequences. This indicates that these Crenarchaeota are facultatively anaerobic or that there are closely related aerobic and anaerobic Crenarchaeota present. Being able to grow drymoderate soil Crenarchaeota opens great possibilities for future molecular and physiological studies as we aim to understand the physiology and the ecological functions of these organisms.

Euryarchaeota in Mycorrhizospheres

In Scots pine both the ectomycorrhizas and mycorrhizal fungal external mycelia not only harboured I.1c Crenarchaeota, but also Euryarchaeota similar to Halobacterium spp. according to their 16S rRNA gene sequence. In this experiment four out of five different species of ectomycorrhizal fungi consistently harboured a steady population of these Euryarchaeota. Euryarchaeotal 16S rRNA genes were never found in the fine roots of Scots pine or the soil outside the mycorrhizosphere. These Halobacterium like related Euryarchaeota from Scots pine mycorrhizas could be grown in anaerobic cultures on yeast extract medium in low saline conditions. Encountering Euryarchaeota with very close 16SrRNA gene sequence similarity to genus Halobacterium in mycorrhizosphere and furthermore being able to grow them in non-saline enrichment conditions contradicts all former knowledge of these usually hyperhalophilic organisms. All previously isolated strains of this genus perish without at least 1 M(5.8%) NaCl in their environment. Whether the physiology of these Euryarchaeota really has any resemblance to the other Halobacterium spp. and how they make a living in the mycorrhizosphere remains to be investigated in future studies. The results gained thus far indicate that they are restricted to the fungal habitats. They may even live inside the ectomycorrhizal fungal hyphae as a common ectomycorrhizal fungus Laccaria bicolor has already been shown to contain intracellular bacteria.

Euryarchaeotal 16S rRNA gene sequences clustering with Methanolobus have been detected from methane producing anaerobic enrichments made from Scots pine ectomycorrhizas. Methanolobus belong to Methanosarcinales, which have shown some tolerance to oxygen and are one of the main groups found in rice field soil where aerobic and anaerobic periods alternate. Methanomicrobiales, the sister taxa to Methanosarcinales, are also tolerant to some soil aeration and have been reported from Spitsbergen wet peat lands. They were, however, not detected in soils which were wet only during spring melt. One reason for not detecting methanogenic Euryarchaeota in the drier habitats in the study by Hoj et al. could be the lack of such hot spots for methanogenic archaea such as ectomycorrhizas as the vegetation was dominated by mosses and even in the driest habitat the only potentially ectomycorrhizal plant was Salix polaris. Different 16S rRNA gene sequences of methanogenic archaea were also detected directly in the roots and mycorrhizosphere of both Scots pine and Silver birch growing in Finnish boreal forest humus. Ectomycorrhizas were the most potential sources for microbes possessing methanogenic activity as ectomycorrhizal in oculum produced roughly ten times as much methane per volume as in oculum from soil outside mycorrhizosphere. Although methane emissions have been reported from forest soils prior to the ectomycorrhizal studies 16S rRNA genes of methanogenic archaea have not been detected from the forest soil samples. It may be that ectomycorrhizas provide a specific niche for methanogenic archaea perhaps possessing areas of lower oxygen availability due to microbial respiration and tuberculate structure. There have been a few 16SrRNA genes of methanogenic archaea as well as tell tale methanogenic activity found in many types of soils but due to the technical difficulties and the apparently rather low numbers of these organisms in dry soils they are difficult to detect.

Manipulation Experiments

In general different manipulations of native soil appear to decrease the diversity and detection frequency of archaea in different soils especially in comparison to bacteria. Heavy metal treatment decreased the percentage of soil archaea from roughly 1.3% of all prokaryotes to below detection level. Reduction in soil archaeal proportion was observed also after heavy metal contaminated sewage sludge treatment. However, when cadmium was separately studied in a forest soil microcosm experiment, it had no effect on the soil crenarchaeotal communities.

In originally low nutrient sandy soil fertilization decreased the proportion of archaea of all prokaryotes. In contrast, methane production as well as novel euryarchaeotal 16S rRNA gene sequences were reported from grassland soils when they were amended with nitrogen fertilizer and kept under intensive sheep grazing. However, it is not clear in the study, if the samples were taken during the time the sheep were already grazing the former site and thus likely adding organic manure and euryarchaeotal (putative methanogenic) in oculum to the soil. This seems likely as pH manipulation, ammonium or urine amendments did not change the archaeal communities of similar grassland soils.

Some archaea appear also to be sensitive to freezing. Pesaro et al. conducted an experiment, where they used sieved calcareous agricultural soil from Rhône valley, which was incubated for upto 42 days. One of the tested treatments was freezing in -20° C. Group I.1a Crenarchaeota were more sensitive to freezing than group I.1b Crenarchaeota. Whether the Crenarchaeota of the former group are sensitive themselves or associated with more freeze sensitive organisms such as fungi or protozoa.

Clear-cutting of mixed coniferous boreal forest changed the soil archaeal community composition. The archaea found in untreated standing forest and clear-cut forest soil were all I.1c group Crenarchaeota. Nevertheless, the differently managed sites only shared two similar sequences. Elevated atmospheric CO_2 increased the abundance of trembling aspen soil Crenarchaeota by 40% as well as Homobasidiomycetes of which the ectomycorrhizal fungus Inocybe sp. was a dominating representative in the samples. This result again points towards the correlation between plant carbon allocation, mycorrhizal fungi and archaea.

Functions of the Soil Archaea

Several soil Crenarchaeota of the group I.1b have been shown to have genes encoding proteins related to subunits of ammonia monooxygenases. These are the central ammonia oxidation enzymes of bacterial nitrifiers. This indicates that soil Crenarchaeota could use ammonia as energy source. Crenarchaeotal AmoA genes have been estimated to be up to 3,000-fold more abundant than bacterial AmoA genes in soils. Thus it may be that archaea are actually more important ammonia-oxidising organisms in soil ecosystems than bacteria. In a fertilization treatment experiment archaeal AmoA genes were always more frequent than bacterial AmoA genes (1.02–12.36x) regardless of the soil treatment. Community structure of soil archaeal ammonia oxidisers has also been shown to react strongly to temperature. Many ectomycorrhizal fungi have very high nitrogen uptake efficiency. As boreal forest soils are dominatingly pervaded with the external mycelia of these fungi, they may out compete the archaeal ammonia oxidisers. This competition could be one explanation for the lack of the commonly found I.1b Crenarchaeota in boreal forest soils. Ammonia oxidation genes have not been detected from the I.1c Crenarchaeota. I.1c Crenarchaeota are phylogenetically quite distant from the other Crenarchaeota and they might not participate in the oxidation of ammonia.

Instead of participating in ammonia oxidation the I.1c Crenarchaeota may have functions in the anaerobic and aerobic cycling of C-1 compounds in the mycorrhizosphere and humus. They have been found only in organic soils with an established plant cover. This may imply that they are either directly supported by the plants and connected mycorrhizal fungi or that they utilize substrates released from the organic matter. In both aerobic and anaerobic enrichment cultures numerous I.1c crenarchaeotal 16S rRNA genes could be found when methane was used as the only energy source. In aerobic cultures I.1c Crenarchaeota could be enriched on methanol as well. Isotopic experiments by West and Schmidt also indicate that Crenarchaeota in alpine soil systems may be involved in oxidation of C-1 compounds including methanol.

Some Crenarchaeota have been found to have the genes commonly found in symbiotic nitrogen fixing bacteria (electron transport chain Fix ABCX, nitrogen regulatory protein PII) needed for nitrogen fixation. A membrane-bound nitrate reductase, which is indicative of denitrification potential, has been described from archaea, but this was a hyperthermophile Pyrobaculum aerophilum.

Nitrogen Fixation

Nitrogen is an important macronutrient because it is part of nucleic acids and proteins. Atmospheric nitrogen, which is the diatomic molecule N_2, or dinitrogen, is the largest pool of nitrogen in terrestrial ecosystems. However, plants cannot take advantage of this nitrogen because they do not have the necessary enzymes to convert it into biologically useful forms. However, nitrogen can be "fixed." It can be converted to ammonia (NH_3) through biological, physical, or chemical processes. Biological nitrogen fixation (BNF), the conversion of atmospheric nitrogen (N_2) into ammonia (NH_3), is exclusively carried out by prokaryotes, such as soil bacteria or cyanobacteria. Biological processes contribute 65 percent of the nitrogen used in agriculture.

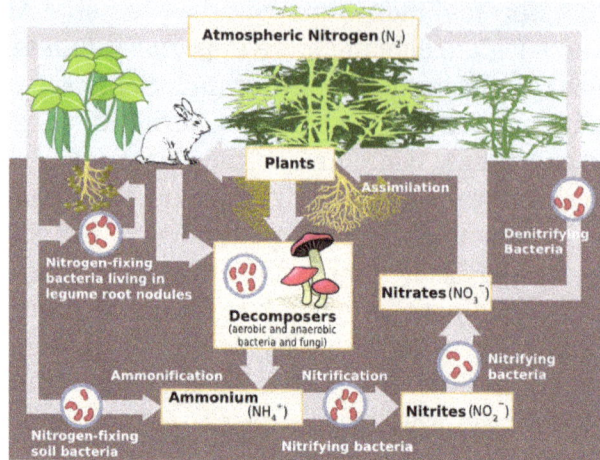

Diagram of the Nitrogen Cycle: Schematic representation of the nitrogen cycle.
Abiotic nitrogen fixation has been omitted.

The most important source of BNF is the symbiotic interaction between soil bacteria and legume plants, including many crops important to humans. The NH_3 resulting from fixation can be transported into plant tissue and incorporated into amino acids, which are then made into plant proteins. Some legume seeds, such as soybeans and peanuts, contain high levels of protein and are among the most important agricultural sources of protein in the world.

Nitrogen fixation in crops: Some common edible legumes, such as (a) peanuts, (b) beans, and
(c) chickpeas, are able to interact symbiotically with soil bacteria that fix nitrogen.

Soil bacteria, collectively called rhizobia, symbiotically interact with legume roots to form specialized structures called nodules in which nitrogen fixation takes place. This process entails the reduction of atmospheric nitrogen to ammonia by means of the enzyme nitrogenase. Therefore, using rhizobia is a natural and environmentally-friendly way to fertilize plants as opposed to chemical fertilization that uses a non-renewable resource, such as natural gas. Through symbiotic nitrogen fixation, the plant benefits from using an endless source of nitrogen from the atmosphere. The process simultaneously contributes to soil fertility because the plant root system leaves behind some of the biologically available nitrogen. As in any symbiosis, both organisms benefit from the interaction: the plant obtains ammonia and bacteria obtain carbon compounds generated through photosynthesis, as well as a protected niche in which to grow.

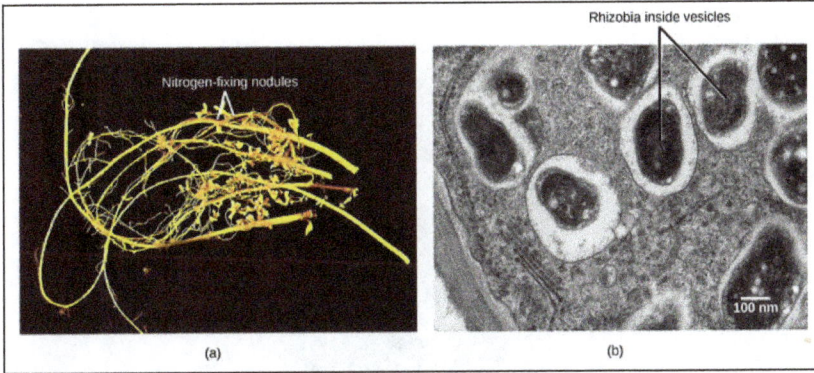

Rhizobia: Soybean roots contain (a) nitrogen-fixing nodules. Cells within the nodules are infected with Bradyrhyzobium japonicum, a rhizobia or "root-loving" bacterium. The bacteria are encased in (b) vesicles inside the cell, as can be seen in this transmission electron micrograph.

- Diatomic nitrogen is abundant in the atmosphere and soil, but plants are unable to use it because they do not have the necessary enzyme, nitrogenase, to convert it into a form that they can use to make proteins.

- Soil bacteria, or rhizobia, are able to perform biological nitrogen fixation in which atmospheric nitrogen gas (N_2) is converted into the ammonia (NH_3) that plants are able to use to synthesize proteins.

- Both the plants and the bacteria benefit from the process of nitrogen fixation; the plant obtains the nitrogen it needs to synthesize proteins, while the bacteria obtain carbon from the plant and a secure environment to inhabit within the plant roots.

Harmful Bacteria in Soil

Harmful bacteria found within soil can cause potential harm to humans, plants and trees. Some forms of bacteria can produce poisonous toxins, which can be fatal if the spores of such bacteria are inhaled, ingested or transferred through a wound.

Bacillus Species

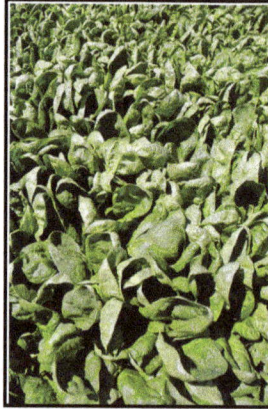

Food contaminated with B. cereus can result in food poisoning.

There are a few variety of Bacillus. Bacillus cereus is a bacteria commonly found in soil. B. cereus is capable of withstanding extreme conditions, such as heat. Food grown in soil containing B. cereus can become susceptible to contamination. It is also possible to inhale aggravated B. cereus spores, or have spores enter broken skin when you don't wear gloves while gardening. According to the Textbook of Bacteriology, B. cereus contains three types of enterotoxins. Enterotoxins are toxins produced by bacteria and are responsible for causing the vomiting and diarrhea associated with food poisoning.

Crown Gall Disease

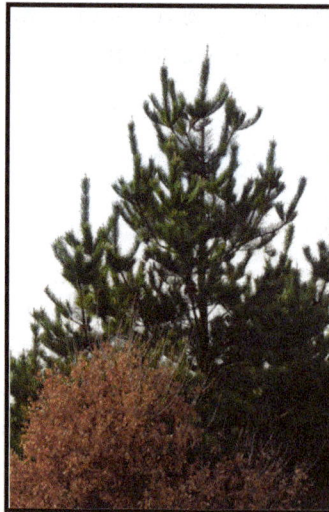

Crown gall affects plants and trees.

Agrobacterium tumefaciens is a form of bacteria that causes disease in plant tissue. If A. tumefaciens enters a healthy tree or plant through the root or stem from the soil, the bacterium will parasitize the tree or plant. The host of A. tumifaciens will succumb to tumor development and changes in plant metabolism. Tumors can begin as white callused tissue on the tree or plant. To prevent crown gall disease, it is important for plant life to be sustained outside of contaminated soil.

Anthrax Development

B. anthracis can survive in soil for years.

Anthrax is caused by the bacterium Bacillus anthracis. B. anthracis can survive for years within soil. When the bacterium produces spores, the potential for contamination becomes possible. Spores can be disturbed during gardening. According to the Directors of Health Promotion and Education, the inhalation of spores from contaminated soil can result in illness. Anthrax is also responsible for producing a toxin that can result in skin ulcers, respiratory distress, fever, vomiting, diarrhea, nausea and possible death.

Anaerobic Bacteria

C. perfringens is most commonly found in soil and water.

Anaerobic bacteria don't require oxygen to survive. One particular species, Clostridium perfringens can be found virtually everywhere. The bacteria can be found in the intestines of humans and animals. However, the bacteria is predominantly found in soil and water. C. perfringens is one of the most common bacterium responsible for food-borne illnesses. If C. perfringens is exposed to extreme temperatures, such as heat, the bacterium will sporulate, or form new spores. The new spores are extremely resistent; which can make outbreaks of illness possible.

Pathogenic Bacteria in Soil

Pathogens are microbes that cause disease in their hosts. While plants usually have more to fear from soilborne pathogenic fungi than from soilborne bacterial diseases, a variety of pathogenic bacteria that live in soil cause disease in plants as well. There are too many species of soil plant pathogens to list them all, but the following are some of the most important or common. A few species of bacteria in the soil can prove dangerous or even deadly to humans as well.

Clostridium Botulinum

C. botulinum is a gram-positive bacterium (i.e. a bacterium that tests positive on the Gram staining test) that forms hardy, durable spores. When the spores find the right conditions, they germinate and begin to grow. They can produce a lethal protein called botulinum toxin, which acts on the nervous system in humans and other mammals to block release of a key neurotransmitter called acetylcholine. C. botulinum spores are found in soil and water in many regions around the world. They are anaerobic and only grow in the absence of oxygen; acidic conditions inhibit their growth. Although C. botulinum sometimes causes fatal cases of a food-borne illness called botulism, the powerful toxin produced by the bacterium has also become useful for paralyzing muscle tissue in humans as part of various medical or cosmetic procedures. When used as a therapeutic agent, botulinum toxin is known by its trade name, Botox.

Erwinia Carotovora

E. carotovora is a highly destructive pathogen that causes disease in a wide range of plants, including potatoes and most other fleshy vegetables. This gram-negative, rod-shaped bacterium can invade the crops in the field or while in storage. The subspecies E. carotovora carotovora survives well in soil and surface waters; it can grow in the root zones of many other nonhost crop or even weed species, invading potatoes or other vulnerable crops when opportunity offers. Common potato diseases caused by E. carotovora include blackleg, aerial stem rot, and tuber soft rot.

Streptomyces Scabies

Streptomyces scabies are gram-positive bacteria that together with two other Streptomyces species (S. acidiscabies and S. turgidiscabies) can infect potatoes and other root crops like beets, radishes, turnips, carrots and parsnips. It's unusual in that as it grows, the cells form filaments similar to those formed by fungi albeit far smaller. Filaments of S. scabies can then break off to form spores. The bacteria spread by invading young, fresh tissues and gaining entry to older tissue by way of wounds or natural openings.

Xanthomonas

Xanthomonas is a genus that includes many important plant pathogens. Although they do not persist for very long in the soil, some species can survive in the soil over the winter and infect a new crop in the spring. Like all other proteobacteria, they are gram-negative. The species X. campestris causes black rot in a wide variety of plants and is perhaps the most important member of the genus. Ironically, the same bacterium is grown commercially to produce xanthan gum, a common food additive.

References

- Introduction-to-actinobacteria, actinobacteria-basics, biotechnological-applications: intechopen.com, Retrieved 24 March , 2019

- Rhizobium: microscopemaster.com, Retrieved 21 March, 2019

- Actinomycetes: maximumyield.com, Retrieved 12 April, 2019

- Actinomycetes-economic-importance-and-reproduction, bacteria: biologydiscussion.com, Retritrieved 23 June, 2019

- Root-and-Bacteria-Interactions, Nitrogen-Fixation, Nutritional-Adaptations-of-Plants, Soil-and-Plant-Nutrition, General-Biology: libretexts.org, Retrieved 15 May, 2019

- Harmful-bacteria-in-soil: hunker.com, Retrieved 20 July, 2019

- Pathogenic-bacteria-soil: ehow.com, Retrieved 25 August, 2019

Soil Fungus and Virus

Fungi play an important role in the soil in various capacities such as food sources for larger organisms and beneficial symbiotic relationships with plants. Viruses also play a significant role in the soil due to their ability to transfer genes from host to host. The topics elaborated in this chapter will help in gaining a better perspective about the different fungi and viruses found in the soil as well as the roles which they play.

Soil Fungus

Fungi are an important part of the microbial ecology. The majority of fungi decompose the lignin and the hard-to-digest soil organic matter, but some fungi consume simple sugars. Fungi dominate in low pH or slightly acidic soils where soils tend to be undisturbed. Fungi break down the organic residues so that many different types of microbes can start to decompose and process the residues into usable products.

Approximately 80 to 90 percent of all plants form symbiotic mycorrhizae fungi relationships by forming hyphae networks. The hyphae are about 1/60 the diameter of most plant root hairs and assist the plant in acquiring nitrogen, phosphorus, micronutrients and water in exchange for sugar produced by the plant. This mutually beneficial relationship is called a mycorrhizae network. Figure shows soil fungus forming mycorrhizae networks.

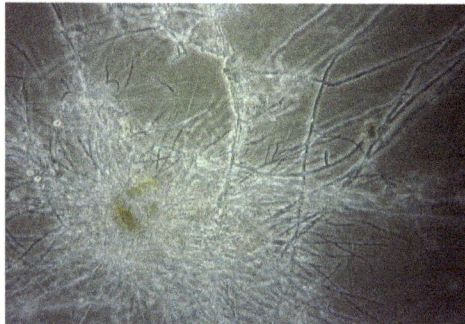

Soil fungus form mycorrhizae networks like a spider web to explore the soil profile for soil nutrients.

Hyphae interact with soil particles, roots, and rocks forming a filamentous body that promotes foraging for soil nutrients. These networks release enzymes into the soil and break down complex molecules that the filaments then reabsorb. Fungi act like natural recycling bins, reabsorbing and redistributing soil nutrients back to plant roots. Most hyphae are either pure white or yellow and are often misidentified as plant hair roots.

Mycorrhizal fungi are usually white or yellow while the root
at the top is a light brown or tan color.

The rhizosphere is an area next to the root dominated by soil microbes where many chemical and biochemical process occur. Soil fungi make up 10 to 30 percent of the soil rhizosphere. Generally there are fewer individual fungi than bacteria but fungi dominate the total biomass due to their larger size in a healthy soil. Fungi biomass in the soil ranges from the equivalent of two to six cows in a healthy soil or 1,100 to 11,000 pounds of biomass.

Fungi prefer slightly acidic conditions, low disturbance soils, perennial plants, internal nutrient sources directly from the plant, and highly stable forms of organic residues with high carbon to nitrogen (C:N) values and slower recycling time. Bacteria dominate in highly disturbed ecosystems with fast nutrient recycling, low C:N values, prefer annual plants, and external nutrient additions outside the plant. Bacteria are single-celled organisms and need a film of water to survive, while fungi are multi-celled organisms that grow rapidly and in great lengths in the soil (feet or meters). This allows fungi to bridge gaps in the soil so as to transport nutrients relatively far distances back to the plants.

Taxonomy and Functional Groups of Fungi

There are at least 70,000 different species of fungi identified but it is estimated that there may be 1.5 million species worldwide. Genetically, fungi evolved a billion years ago and are closely related to plants and animals. Membrane bound organelles present in each cell are similar to those found in insects, plants and animals. Fungi have 80 percent or more of the same genes as humans.

There are four major groups of soil fungus: Zygomycota, Ascomycota, Basidiomycota, and Deuteromycota. Zygomycota are less than 1,000 species and are mostly common bread molds. Ascomycetes have about 30,000 species and are mostly yeasts used in baking. Basidiomycetes include most mushrooms, toadstools and puffballs, while Deuteromycota include the lichens and the mycorrhizal fungus. Fungi are classified as heterotrophs so the carbon source originates from the decomposition of organic compounds or residues.

Decomposers are also called saprophytic fungi which decompose cellulose and lignin in the soil. Sugar fungi called Zygomycetes decompose the simple sugars but most fungi decompose the more recalcitrant or hard-to-decompose organic residues high in cellulose, hemicellulose, lignin or cell walls. Some of the byproducts of this decomposition may turn to humus and remain in the soil for thousands of years.

Pathogenic fungi cause many agricultural root diseases including Phytophthora, Rhizoctonia, Phythium, and Verticullium, and downy mildew Ascomycetes fungi are microscopic in size and dominate in agricultural soils and grassland while the Basidiomycetes have large fruiting bodies or mushrooms that dominate in high residue and forested soil. Some fungi help to control diseases and predators including a nematode trapping fungi that feeds on insects and can be used as biological controls.

Mutualistic mycorrhizal fungi form a beneficial relationship with plants. Ingham states that "mycorrhizae grow within the root cells and are commonly associated with grasses, row crops, vegetables, and shrubs. Some plant species like the Cruciferae family (e.g., cabbage, broccoli, mustard and canola) and the Chenopodiaceae family (e.g., lambsquarters, spinach, beets, and oilseed radish) do not form mycorrhizae associations."

Ecological Plant-Microbe Interactions

The microbes and plants together regulate many soil processes including the carbon cycle and nutrient recycling. The microbial species diversity and the total microbial population determine the ability of plants to obtain soil nutrients like nitrogen, phosphorus and micronutrients. Plant diversity and abundance may change the entire soil ecosystem through the release of root exudates that attract or inhibit the growth of specific organisms. These carbon-rich substances can range from less than 10 percent to as much as 20 percent of a plant's total carbon production. Plants secrete large amounts of specific carbon compounds into the soil to be used by the microbes as a food and energy source and to enhance and improve soil structure. By encouraging certain microbial species to grow, the microbes in return supply nutrients to the plant roots and encourage the microbes to protect the plant from pathogenic microbes. Plants feed, raise and encourage certain microbes just like farmers raise and feed plants and livestock for food and fiber.

Many plants have a preferred fungus to bacteria ratio (F:B ratio). Most vegetable crops prefer more bacteria with a F:B ratio of 0.3 to 0.8 F:B which is optimal for carrots, lettuce, broccoli and cole crops. Tomatoes, corn and wheat like an F:B ratio of 0.8 to 1:1. Lawns prefer a F:B ratio of 0.5 to 1:1 and on forested soils, trees grow better with a F:B ratio of 10:1 where the soil is more acidic. Many conventionally tilled agricultural soils have a F:B ratio of 0.1 to 0.3 with soils that are high in nitrogen, low carbon, neutral pH, and with disturbed soil conditions which promotes weed production. Annual crops prefer lower F:B ratios and perennials prefer a higher F:B ratio. Many plants cultivate certain species of both bacteria and fungus to increase nutrient extraction from the soil.

Fungi benefit most plants by suppressing plant root diseases and fungi promote health-ier plants by attacking plant pathogens with fungal enzymes. Fungi also use antag-onism to reduce competition by producing antibodies, which suppress other micro-organisms from growing. They produce many vitamins which promote plant growth. By competing with other fungus for nutrients, beneficial fungi prevent pathogenic and disease-causing organisms from getting established.

Beneficial fungi along with some bacteria may also form protective webs and nets around roots and even leaves to protect the host plant. The fungus Trichoderma protects plant roots from attack by harmful microorganisms. Fungi also protect plants by supplying a protective sheath to supply both water and phosphorus to the plant roots during droughts.

Hyphae must be in close contact with living or dead organic soil residues to absorb nutrients, so they usually grow in association with other soil microorganisms. They compete aggressively for scarce nutrients, and competition usually results in a succes-sion or change in microbial populations as nutrients are absorbed or depleted. Initial colonizers absorb simple sugars, amino acids, and vitamins from plant parts such as fruits, seeds and vegetables, and are classified as "sugar fungus." The dominance of these fungi is short-lived because waste products accumulate quickly.

Cellulose degraders appear next and they are the most diverse fungi and competitive. For example, degradation of straw with high C:N ratio (80:1) requires that fungus par-asitize or decompose other fungi to obtain nitrogen for growth and enzyme production. Degradation of lignin is stimulated by low nitrogen. Lignin makes up 60 percent of the total mass of humus, but the low number of fungal species that can degrade lignin reduces competition.

Fungi generally reproduce asexually by spores (microscopic parts similar to plant seeds). Spore dispersal occurs in a variety of ways including triggers, jet propulsion, springs, and scents depending on the environment. Fungi spores may live 50 years in an inactive vegetative state as a spore. This allows the fungi to survive and remain via-ble by staying inactive until environmental conditions improve. Microbes spend most of their time hungry and looking for nutrient resources just to survive but fungus may remain viable for years. Most fungi survive by continual growth and translocation of nutrients in wide networks as they search for food or they survive through the produc-tion of resistant spores during times of stress. Natural soils tend to suppress germinat-ing fungal spores, especially when nutrients are limiting.

The longevity of fungi has not been measured in many species but their open-ended growth suggests that they have a longevity measured in millions of years, because they are basically the same organism. For example, fairy fungal rings grow in ever widening circles, much like rings on a tree, and are measured in decades and centuries instead of days and weeks for most microbes. One fungus in Michigan's Upper Peninsula cov-ered 20 hectares or almost 42 acres, connected itself to most of the trees and weighed 10,000 kilograms or 11 ton and was estimated to be over 1,500 years old.

Most soil fungi decompose recalcitrant organic residues high in cellulose and lignin. Fungi carbon use efficiency is about 40–55 percent so they store and recycle more C (10:1 C:N ratio) and less N (10 percent) in their cells than bacteria. While fungi are smaller in number, they equal or exceed the biomass of bacteria due to their greater size. Fungi are more specialized but need a constant food source and grow better under undisturbed soil conditions. Fungus and bacteria recycle soil nutrients and generally have a symbiotic relationship with most plants.

Common Types of Fungi Found in Soil

At least 70,000 distinct species of soil fungi have been identified worldwide. These can be divided taxonomically into four groups: Zygomycota, Ascomycota, Basidiomycota, and Deuteromycota. It is perhaps easier, however, for everyday observers to think of them in terms of their function and metabolic properties.

Saprophytic Fungi

Saprophytic fungi are decomposers. Fungal saprophytes decompose cellulose and lignin found in soil for energy. The metabolic byproducts of this process include carbon dioxide, or CO_2, and small molecules such as organic acids; some of these metabolites can remain in the surrounding soil for thousands of years under the right conditions.

Saprophytic mushrooms can grow in a wide variety of locations because of the widespread nature of the substances from which they derive nourishment. Some saprophytic fungi are called "sugar fungi" because they use the same substrates as do many bacteria.

Mutualistic Fungi

Mutualistic fungi are also called mycorrhizal fungi. These soil fungi colonize plant roots and are termed "mutualistic" because the fungi derive a benefit from the presence of the plants and vice versa. In exchange for carbon atoms from the plant, mycorrhizal fungi help make phosphorus easier for the plants to draw in, and they also bring other soil nutrients, including nitrogen, micronutrients and water to the plants on which they are anchored.

Mutualistic fungi include two major groups. One of these is the ectomycorrhizae, which grow on the surface of plant roots and are frequently seen on or near trees. The second major group, the endomycorrhizae, grow within, rather than on, the plant root cells, and are usually associated with grasses, crops grown in rows, vegetables and everyday shrubs.

Pathogenic Fungi

Also called parasitic fungi, these fungi, in stark contrast to the mutualists, result in reduced plant production or even the death of plants when they colonize roots or other soil organisms. Root-pathogenic fungi cause major annual economic losses in human agricultural endeavors. Many of these fungi, however, can help control diseases when properly deployed. For example, nematode-trapping fungi parasitize disease-causing nematodes, or roundworms, while fungi that derive their energy from insects may be put to use as pest-control agents.

Mycorrhizae

Mycorrhizae (singular: mycorrhiza) are mutualisms formed between fungi and plant roots. The importance of mycorrhizae cannot be overstated; it has been suggested that as many as 95% of all the world's plant species form mycorrhizal relationships with fungi and that in the majority of cases the plant would not survive without them. Mycorrhizae have existed for a very long time and can be demonstrated in the fossilized roots of some of the earliest land plants. They can be found in plants growing in habitats ranging from humid to dry tropics all the way to the far north and south. Some scientists have suggested that plants were only able to move on to land when they had developed mycorrhizal relationships with fungi.

Mycorrhizae are considered to be a mutualistic relationship because both organisms benefit. The fungus receives the products of photosynthesis from the plant and is thus freed from the necesity of finding its own sources of energy. At the same time the fungus grows out into the soil and retrieves nutrients, especially phosphorus and nitrogen, and passes these back to the plant. Numerous experiments have shown that plants without mycorrhizae cannot cope as well with low mineral levels as those that have mycorrhizae.

It is generally believed that mycorrhizae have been re-invented many times over the history of land plants. Many different groups of fungi are involved and the form of the actual fungus-root interface, the mycorrhiza itself, varies greatly. Scientists recognize several distinct types of mycorrhizae and can relate these to particular groups of plants and fungi. The most thoroughly studied of these types are arbuscular mycorrhizae, ectomycorrhizae, ericoid mycorrhizae, arbutoid mycorrhizae and orchid mycorrhizae

Arbuscular Mycorrhizae

Arbuscular mycorrhizae (often called AM) are the most common and widespread of all mycorrhizae and are found in as many as 85%-90% of the world's plant species.

In this association the fungus occurs inside the cells of the plant root as a highly branched shrubby structure called an arbuscule. The figure shows an arbuscule of one of these fungi inside a root cell of a small onion plant. To view this fungus is was necessary to clear all of the contents out of the root cells with hot potassium hydroxide followed by a blue dye called Trypan Blue. The cellulose of the root tissue did not stain very much while the wall material of the fungus, possibly chitosan, stained purple. Although it appears that the arbuscule fills the root cell it actually occurs between the root cell wall and the cell membrane within. The cell membrane fits over the arbuscule like a rubber glove over your hand. Thus the fungus never comes into direct contact with the root cell nucleus, mitochondria or other cell structures. The exchange of nutrients between the two partners, minerals from fungus to plant and sugars from plant to fungus, takes place at the cell membrane-arbuscule interface.

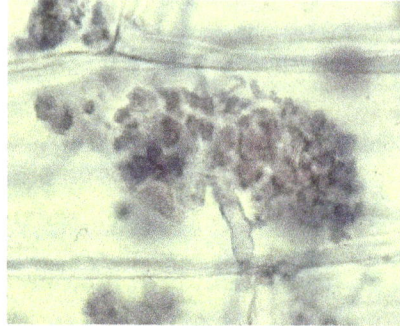

Fungi forming arbuscular mycorrhizae may produce some other structures in addition to arbuscules. Roots harbouring arbuscules may also contain balloon-like structures called vesicles as illustrated above in the root cells. Vesicles are very commonly produced by these fungi, in fact, that they have been referred to as "Vesicular-arbuscular Mycorrhizae (VAM)". However, that name is unnecessary because these fungi nearly always produce arbuscules but may often lack vesicles.

The fungi forming arbuscular mycorrhizae not only occur in roots but also extend out into the surrounding soil where they form a dense network. It is this so-celled extra-radicular phase that is responsible for extracting nutrients from the soil. The hyphal network is relatively long-lived and is able to colonize new roots as they enter its domain. One of the criticisms of soil tilling, a common agricultural practice, is that it breaks up these mycorrhizal networks and prevents germinating seeds and young plants form tapping into existing systems of nutrient uptake.

Reproduction of the fungi forming arbuscular mycorrhizae is by thick-walled spores produced on the extra-radicular hyphae. These are thick-walled and often remain in the soil for long periods. Many are quite large and can be recovered by sieving the soil. Although these spores will germinate and produce hyphae in the laboratory no one has yet succeeded in growing the fungus independently of a root. The below figure shows spores of a species of Glomus produced on a branched extra-radicular hypha.

The fungi able to form arbuscular mycorrhizae are relatively few in number and all are members of their own phylum, the Glomeromycota. Although the root system of a plant might support several species of these fungi the diversity is never great. In fact some species appear to be common throughout the world, in spite of immense climatic and floristic differences.

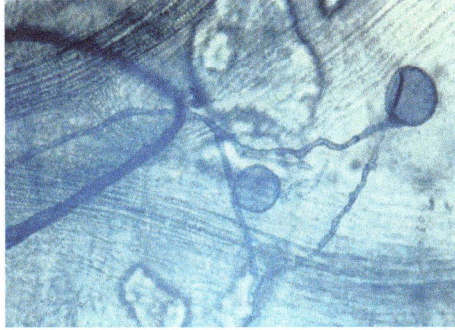

Ectomycorrhizae

Ectomycorrhizae, sometimes abbreviated as EM, are not as widespread in nature as arbuscular mycorrhizae, occurring on the roots of about 5% of the world's plants. The most diagnostic feature of ectomycorrhizae is that the fungus never penetrates the cell walls of the plant and thus the exchange of nutrients must take place not only through the cell membrane but through both plant and fungal cell walls as well.

A young root of paper birch is shown in the illustration at right. It was stained with Trypan Blue as outlined above and shows an older section uncolonized by fungi and thus unstained giving rise to a new branch that has become heavily colonized. The outside of the root is ensheathed in a thick covering of hyphae called a mantle, visible in the picture as a dark blue border around the root. The mantle is outside the root and can often be removed as though it were a shell. It completely isolates the root from its environment so that all substances coming into that root must come from the fungus. Inside the mantle is a single layer of cells forming the epidermis, the outermost tissue of the root. In paper birch and several other species the epidermal cells become elongated and slanted to form a herringbone pattern following fungal colonization. Below the epidermis is a layer of cells forming the cortex of the root. The cortex varies in thickness according to the species of plant, ranging from one to several cells deep. The colonizing fungus grows between the cells of the epidermis, and often the cortex as well, and surrounds the individual cells, finally isolating them from one another. This network of enveloping hyphae surrounding epidermal and often cortical cells is called the Hartig net in honour of the nineteenth-century German biologist Robert Hartig who first observed it. The Hartig net is the interface between the fungus and the plant; it is the site of all nutrient and water exchanges between the two associates. Although the Hartig net surrounds the plant cells they do remain in contact with one another through microscopic strands called plasmodesmata. The centre of the root is taken up by the stele, the conducting system of the root. Water and nutrients passed by the fungus to the root are transported throughout the plant via the stele, itself composed to several parts.

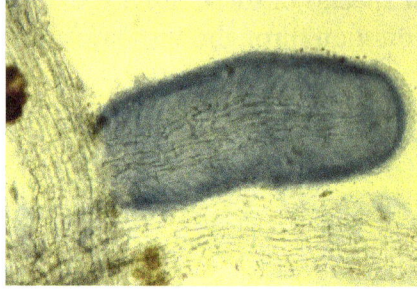

The above figure illustrates these relationships more clearly. It presents a narrow strip across a mycorrhizal root of Populus tremuloides, the trembling aspen. Comparing this picture with the one above of paper birch you can see that the root tip would be at the bottom of the page. At this higher magnification the compactly arranged hyphae of the mantle can be seen as well as the hyphae of the Hartig net surrounding the slanted epidermal cells. The cortex can be seen to be made up of more or less round cells while the stele is clearly formed of elongated cells. In the aspen and birch roots the Hartig net extends only into the epidermis, but in other species, white spruce for example, the cortex may also become involved. The location of the fungus within the root is under the control of the plant, which uses toxic phenolic compounds to limit growth of the hyphae. This can be observed in laboratory experiments where the same species of fungus is introduced into the roots of two species of plants; in one it may extend down into the cortex while in the other it may be restricted to the epidermis.

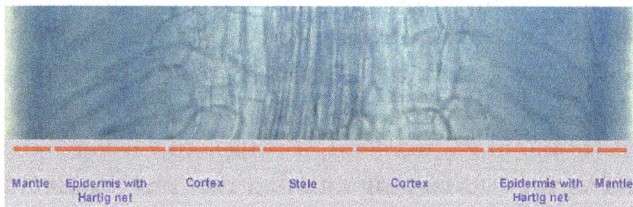

Unlike the fungi forming arbuscular mycorrhizae, those forming ectomycorrhizae are exceedingly diverse. Several thousand species of these fungi are known and many are undoubtedly still waiting to be discovered. With such a diversity of fungi on hand, the root systems of large ectomycorrhizal plants may support dozens of fungal species. Moreover, at least once, and possibly more times in a year the root grows out from its mycorrhizal sheath and must be recolonized. The figure at far left shows an aspen root growing out through the end of a mycorrhizal sheath, probably produced by the fungus Cenococcum geophilum. Although a little out of focus you can see that the emerging root tip is uncoloured and unstained, indicating that it was yet to be colonized. The figure in the right panel shows an aspen root that has had three such growth spurts: the first of these resulting in colonization by a lightly staining but orginally colourless fungus, followed by two additional colonizations by a brown fungus. It appears that the root was orginally colonized by the blue-staining species but that this was replaced by another species following a growth spurt. The third growth spurt simply resulted in recolonization by the brown species. These events suggest that as a root grows it may continue to associate with one species of fungus but it also has the opportunity to change partners. Numerous studies have shown that ectomycorrhizal fungi differ greatly from one another in their physiology and it is tempting to think that as a root extends out through the soil it is able to form an association with a fungus apprpriate to the particular conditions it encounters.

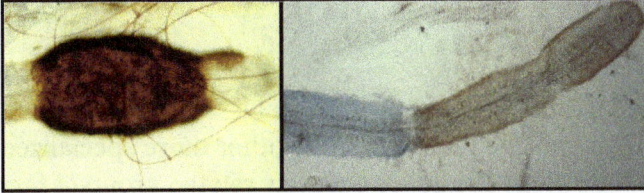

The ability to form ectomycorrhizae is found in many families of fungi, but most commonly among members of the class Agaricomycotina of the Basidiomycota, especially those producing mushrooms and boletes. Most of the larger mushrooms you see in the forest have arisen from the networks of extra-radicular hyphae permiating the soil beneath your feet. The abundance of these mushrooms, their sheer weight and volume, attests to the magnitude of their activities. The energy and chemicals needed to build these mushrooms comes in great part from the trees, suggesting that the advantages a plant gains from mycorrhizae come at a cost.

Most plants exclusively form arbuscular mycorrhizae but there are compelling reasons to focus attention on those having ectomycorrhizae as well. Although a smaller number of species are involved, ectomycorrhizae dominate in the pine, oak, birch, willow, walnut and several other families. In the tropics these include the dipterocarps and large woody legumes. In New Brunswick our extensive forests of spruce, fir, white pine, birch and poplar support immense continuous networks of ectomycorrhizal fungi. Without these fungi our forests as we know them would not exist. Thus the ecological and economic importance of ectomycorrhizae cannot be overestimated.

Many biologists have noted the major differences between tropical and temperate forests and have attempted to relate these to dominance by certain mycorrhizal types. The pictures above illustrate two such forests; at left a tropical rain forest in northern Costa Rica and at right a forest near Schefferville, Quebec. The Costa Rican forest is dense and made up of a great variety of tree species. You might walk some distance through this forest before encountering two individual trees of the same species. Biodiversity here, including the trees, seems to be high. On the other hand the Quebec forest appears to have only one kind of tree. Closer examination would reveal some four or five species but hardly more. If you started walking away from the base of a spruce tree it wouldn't be long before you encountered another. Biodiversity here seems to be very low. Curiously, fungal biodiversity in these forests takes another form, at least when it comes to mushrooms. Few trees in the Costa Rican forest are able to form ectomycor-rhizae while in the Quebec forest all the trees form ectomycorrhizae. As a result of this the Quebec forest will have a great variety of large mushrooms while the tropical one will support a lessor variety of mostly small mushrooms.

Other kinds of mycorrhizae Although arbuscular and ectomycorrhizae account for most instances of mycorrhizae there are some other more specialized types. These are all restricted to a single plant family, or a closely related group of families, and are not as commonly encountered as the two main types. However, in New Brunswick we are fortunate to have these other types of mycorrhizae and can observe them during the active growing season.

The most widespread of the "minor" groups of mycorrhize are ericoid mycorrhizae, so named because they are restricted to the Ericaceae, the heath family, and some closely related families. The Ericaceae are familiar to New Brunswickers in the form of blue-berries, cranberries, sheep laurel, rhododendron, etc. Crowberries, in the related family Empetraceae, also have ericoid mycorrhizae. Figute above at right shows a root of Lab-rador tea, another member of the Ericaceae common throughout New Brunswick. Com-pared with the birch and aspen roots those of Labrador tea appear to be lacking some tissues, but in fact everything is there. These roots are exceedingly thin and often called "hair roots". If you attempt to pull one of these plants out of the soil without loosening the roots first with a shovel or trowel you will leave all of the hair roots in the ground. In the picture there appears to be only a cortex and a stele but there is also an inconspic-uous epidermis that didn't show up very well in the photograph. The cortex is only one cell thick and these cells are packed with fungal hyphae. In many ericoid mycorrhizae the fungus may actually outweight the root. Such a root is little more than a receptacle for holding a fungus. Although the hyphae appear to fill the cortical cells they are out-side the cell membrane, which is intricately wrapped around them. Nutrient exchange between the fungus and plant takes place at this interface and is extremely efficient.

The plants forming ericoid mycorrhizae are highly adapted to this association and, not surprisingly, so are the fungi. These fungi, most commonly the inoperculate discomy-cete Rhizoscyphus ericae, have highly developed abilities to extract nutrients from nu-trient-poor soils. Hence members of the Ericaceae are able to grow in such places as bogs and sandy areas that would be unavailable to most plants.

Arbutoid mycorrhizae also occur in the Ericaceae but only on a few species, mainly those of Arctostaphylos and Arbutus. Arctostaphylos uva-ursi or bearberry is the only representation in New Brunswick and it is not very common, although it is abundant to the north and west of us. Arbutoid mycorrhizae are similar to ericoid mycorrhizae in having the cortical cells filled with hyphae but differ in having a mantle as in ectomycorrhizae. The fungi involved are apparently ones that also commonly form ectomycorrhizae.

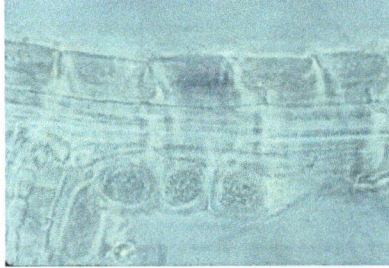

Monotropoid mycorrhizae resemble arbutoid mycorrhizae in many ways but differ in the appearance of the roots and in the nature of the mutualism. In the pictures above, all of Indian pipe, Monotropa uniflora, we see at left the above-ground part of the entirely white plant. The picture in the middle shows the base of the stems, including the underground root mass, and the one at right provides an enlarged view of the roots. Indian pipe is common throughout Canada and lives as a parasite. Until fairly recently it was believed that this plant attached itself to the roots of other plants where it would absorb their nutrients. It's lack of chlorophyll suggested that it had no existance independently of its plant host. More recent studies have demonstrated that Indian pipe and some other related plants actually tap into the extra-radicular hyphae of fungi forming mycorrhizae with neighbouring trees and encorporate these into its own dense, concentrated mycorrhizal system. From these mycorrhizae the plant can drain sugars and all the other nutrients it requires. From this we see that Indian pipe is not really a parasite of plants but of mycorrhizal fungi.

Orchids form their own kind of mycorrhizae, often for complete nutrition including sugars. Many orchids are without chlorophyll when young. Others lack it throughout their lives. The use of mycorrhizae to offset this lack of chlorophyll is similar to the situation in Indian pipe. The mycorrhizae themselves are similar to ericoid mycorrhizae, although the roots are larger and the colonization less extensive. The fungi involved in orchid mycorrhizae are interesting in that they are close relatives of fungi causing plant disease. Mycorrhizal fungi in other groups are not closely related to parasitic groups and would present no threat to the plant.

Role of Soil Fungus

Benefits of Soil Mycorrhizal Fungus

Fungal hyphae have advantages over bacteria in some soil environments. Under dry conditions, fungi may bridge gaps between pockets of moisture and continue to survive and grow, even when soil moisture is too low for most bacteria to be active. Fungi are able to bring nitrogen up from the soil, allowing them to decompose surface residue which is often low in nitrogen.

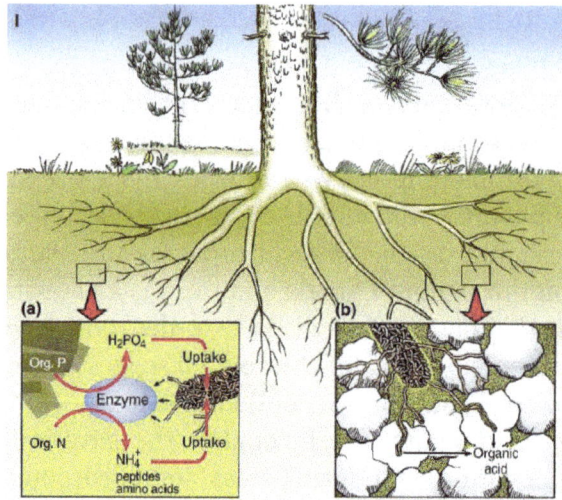

Fungus hyphal filaments translocate and store deficient nutrients to distant parts of the soil where nutrients may be lacking, allowing reproduction and growth to continue. The plant supplies simple sugars to the fungus while fungus supply N, P, other nutrients and possibly water to the plant. As much as 20% of the total carbon assimilated by the plant may be transferred to the fungal partner. The mycorrhizal network of fungi hyphae greatly increases the surface area for plant roots and efficiently transport nutrients back to the plant. Mycorrihzal networks explore up to 20% of the soil volume due to their smaller size compared to only 1% of the soil volume for a typical plant root hair. However, when excess nutrients like N and P are supplied by commercial fertilizer to plant roots, the AM fungi stop working. Tillage also decreases the effectiveness of the AM fungus by destroying the mycorrhizal network associated with plant roots.

Fungal hyphae physically bind soil particles together, creating stable macroaggregates (>250 µm) that help increase water infiltration and soil water holding capacity. Typically there are from 1 to 20 meters of AM hyphae in each gram of soil (as much as 5 miles of AM hyphae in a pound of soil).

AM fungi produce a sticky substance call glomalin. Glomalin is an amino polysaccharide composed of sugars from the plant root and protein from the fungus forming a glycoprotein. Glomalin surrounds the microaggregate soil particles and glues them to-gether to form macroaggregate soil particles. Polysaccharides like glomalin enhance good soil tilth and good soil structure. Some inoculums of mycorrhizal fungi are com-mercially available and may be added to the soil at planting time.

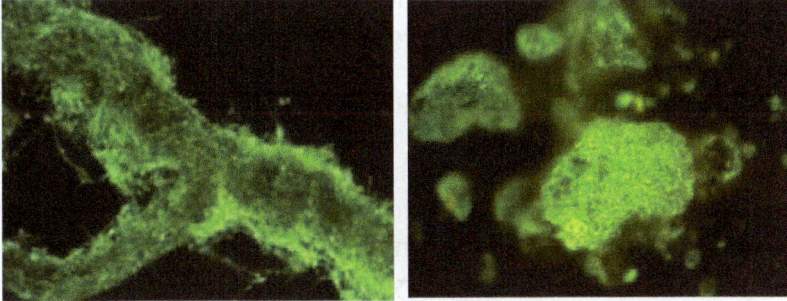

Glomalin surrounding a root heavily infected with mycorrhizal fungi and soil macroaggregates surrounded by glomalin.

Conventional tilled soils are dominated by bacteria, which do not produce glomalin. Tillage disrupts and breaks down the macroaggregates into microaggregates and re-sults in denser, more compacted soils, lacking soil structure. Tilled soils lack soil organ-ic matter (SOM) and the ideal soil habitat to feed and keep large numbers of beneficial fungal populations healthy. Fungus need well-aerated soils and cannot tolerate satu-rated or anaerobic (lack of oxygen) soil conditions that occur under tilled compacted soils. Tillage also injects excess oxygen into the soil, stimulating bacteria populations to increase, and the expanding bacteria populations consume active carbon (polysaccha-rides and glomalin) for food.

In a good soil, glomalin may represent 1–5% of the total carbon in the soil. Glomalin is 30% carbon, 1–2% nitrogen, and up to 5% iron, which gives it a reddish soil color. Broad spectrum fungicides are toxic to mycorrhizal fungi populations. So tillage, excess commercial fertilizer, pesticides, short crop rotations, and long fallow periods tend to decrease fungal populations and decrease the production of glomalin in the soil.

Some plant species like the Cruciferae family (e.g., broccoli, mustard), and the Cheno-podiaceae family (e.g. lambsquarters, spinach, beets) do not form mycorrhizal associ-ations. The level of dependency on AM fungus varies greatly among varieties of some crops, including wheat and corn; however, greater than 90% of plants have an associ-ation with AM fungus.

Ecological Plant-Microbe Interactions

Plant-microbe interactions in the rhizosphere are responsible for a number of soil pro-cesses that include carbon sequestration, ecosystem services, and nutrient recycling. The composition and quantity of microbes in the soil influence the ability of plants to obtain nitrogen and other nutrients. Plant can influence these ecosystem changes through the deposition of root exudates or carbon rich substances into the soil to attract or inhibit the growth of specific organisms. These carbon rich substances can range from less than 10% to as much as 44% of a plant's total carbon production. Soil microbes utilize this abundant carbon source, implying that the plant's selective secretion of specific compounds between plants and microbes may encourage beneficial symbiotic and protective relationships against pathogenic microbes. Plants basically feed, raise, and encourage certain microbes just like farmers raise and feed plants and livestock for food and fiber.

Fungi benefit most plants by suppressing plant root diseases and promoting healthier plants by attacking plant pathogens with fungal enzymes. Fungi enhance and cultivate good bacteria, especially Rhizobia bacteria for nitrogen fixation, which help legume plants grow. Fungi aid plants in N and P soil extraction and help in drought stress to supply more water to plants through the fungal hyphal network and by providing a protective shealth around plant roots under dry conditions. During drought stress, as soils dry down, P becomes limiting even in soils with high P availability. Fungi protect the plants by supplying a protective sheath to supply both water and P to the plant roots.

Mycorrhizal fungus may have over 5 miles of hyphae in a pound of soil. The hyphae are usually white or yellow or tan. The root at the top is tan or light brown color

When environmental conditions decline, fungus form spores, which store large amounts of nutrients to allow the fungus to survive until environmental conditions recover. Fungus with specialized nutrient requirements are more likely to decline or die out when a non-host plant is included in a crop rotation while a fungus with a wide host range will survive and possibly flourish. Soil fungi often have enough resiliency to convert from one resting stage to another when nutrient levels are low. These patterns are important in agricultural management systems that involve fallow periods between cropping cycles.

Fungal hyphae must be in close contact with living or dead organic soil residues to absorb nutrients, so they usually grow in association with other soil microorganisms. They compete aggressively for scarce nutrients, and competition usually results in a succession or change in microbial composition as nutrients are absorbed or depleted. Initial colonizers absorb simple sugars, amino acids, and vitamins from plant parts such as fruits, seeds, and vegetables and are classified as "sugar fungus." The dominance of these fungi is short-lived because waste products accumulate quickly Cellulose degraders appear next and they are the most diverse and competitive. Degradation of straw has a high C:N ratio (80:1) and requires that fungus parasitize or decompose other fungi to obtain nitrogen for growth and enzyme production. Degradation of lignin is stimulated by low nitrogen. Lignin makesup 60% of the total mass of humus, but the low number of fungal species that can degrade lignin reduces competition. Fungi also use antagonism to reduce competition by producing antibodies, which suppress other microorganisms from growing.

Role of Soil Fungus

Fungi perform important services related to water dynamics, nutrient cycling, and disease suppression. Along with bacteria, fungi are important as decomposers in the soil food web, converting hard to digest organic material into usable forms. In no-till, fungal population dominate the soil food web (although they are less in number than the bacteria). Fungi have 40–55% carbon use efficiency so they store and recycle more carbon (C) compared to bacteria. Bacteria are less efficient at retaining C and release more into the air as carbon dioxide.

Fungi have higher C content (10:1 C:N ratio) and less nitrogen (N=10%) in their cells than bacteria. Fungus help recycle both N and phosphorus (P) to plants. Due to their smaller size and much greater surface area, fungus can efficiently scavenge for N and P better than plant root hairs and greatly increase the plant root nutrient extraction efficiency. Many plants cultivate certain species of both bacteria and fungus to increase nutrient extraction from the soil.

Genetically, fungi are closely related to plants and animals. Membrane bound organelles present in each cell are similar to insects, plants, and animals. They evolved about a billion years ago and are equal in rank to plants and animals. In fact, fungi have 80% or more of the same genes as humans. They generally reproduce by spores (microscopic parts similar to plant seeds). The longevity of fungus has not been measured in many species but their openended growth suggests that they have a longevity measured in millions of years, because they are basically the same organism. For example, fairy fungal rings grow in ever widening circles, much like rings on a tree, and are measured in decades and centuries instead of days and weeks for most microbes.

Soil Virus

Soil viruses are of great importance as they may influence the ecology of soil biological communities through both an ability to transfer genes from host to host and as a potential cause of microbial mortality. Consequently, viruses are major players in global geochemical cycles, influencing the turnover and concentration of nutrients and gases.

Despite this importance the area of soil virology is understudied. To explore the role of the viruses in plant health and soil quality, we are studying virus diversity and abundance in different geographic areas (ecosystems) using classical methods of virus purification, electron microscopy and next generation sequencing (metagenomic studies).

We have recently isolated and characterised virus-like particles from soils in the Dundee area of Scotland. Different virus morphotypes including tailed, polyhedral (spherical), rod-shaped, filamentous and bacilliform particles were detected in the soil samples. We have also characterised viruses from different soil functional domains surrounding wheat roots (rhizosheath and rhizosphere), and bulk soil.

We have found that viruses are highly abundant in all the domains. Moreover, in spite of the differences in the abundance of bacterial communities in these domains, no significant variations in viral population structure in terms of morphology and abundance were found. Typically there were $\sim 1.1 - 1.2 \times 10^9$ virions g-1 dry weight of soil, revealing remarkable differences in virus-to-bacteria ratios in domains close to roots ($\sim 0.28 - 0.29$) compared to bulk soil (~ 4.8).

Soils probably harbour many absolutely novel viral species that together may represent a large reservoir of genetic diversity. For example, we have recently isolated a novel bacteriophage from Antarctic soils. The phage, that we have named SpaA1, is morphologically similar to phages of the family Siphoviridae. The 42,784 bp genome of the phage encodes 63 genes which cluster within three regions of the genomes, each of apparently different origin, in a mosaic pattern.

Remarkably, one of the regions contains an almost complete genome of a distinct bacteriophage, MZTP02. These data suggest that MZTP02 can be exchanged between genomes of other bacteriophages, leading to the formation of chimeric genomes. The insertion of a complete phage genome into the genome of another phage has not been described previously and might represent a novel 'fast track' route for virus evolution and horizontal gene transfer.

Studies on Soil Viruses

Many studies have been conducted on viruses in soils from the viewpoints of crop production and epidemiology. With either viewpoint, proliferation, inactivation, survival and the fate of specific viruses were the main concerns, for example, viruses infecting useful and undesirable microorganisms such as rhizobia and plant pathogens in agriculture and infectious to mammals in epidemiology. In other words, the main concerns were focused on the autecology of specific viruses in soils.

Agronomic Concerns

For soil microbiologists, agronomic interests have continuously been the motivation for viral studies in soil since the 1970s. Many studies were conducted to elucidate rhizobiphage populations, the host range of rhizobiphages among Rhizobium spp., and the effects of rhizobiphages on the nodule formation and yield of leguminous plants. El

Didamony and Abdel-Fattah examined the effect of phage infection (lysogenic and lytic to rhizobia) on the growth stimulation of leguminous plants by the arbuscular fungus Glomus mosseae and two Rhizobium leguminosarum bv. viciae strains. Shoot dry weight and nodulation of faba bean plants inoculated with G. mosseae were significantly increased compared with non-mycorrhizal plants in the presence and absence of the phage of the R. leguminosarum strains. The number of nodules produced by the lysogenic rhizobial strain was significantly increased in mycorrhizal plants, but was still less than the number produced by the non-lysogenic strain. The levels of mycorrhizal colonization were markedly enhanced in the presence of the non-lysogenic rhizobial strain.

The effects of viruses on beneficial bacteria and soil-borne plant pathogens have also been extensively studied. In general, phages are abundant in soils where bacterial hosts proliferate, such as rhizobiphages in the fields of legumes. In relation to this context, it is noteworthy that no attention has been paid to soil phages infecting anaerobic bacteria, such as methanogens. As for phages of general soil bacteria, Ashelford et al. observed clear seasonal population dynamics of podophages and their host, Serratia liquefaciens, in sugar beet fields.

Bacterial blight caused by Xanthomonas oryzae pv. oryzae is a destructive disease of rice plants in Asia. It is an air-borne disease and several tadpole-shaped and filamentous phages of the pathogen were isolated from paddy soils and diseased rice leaves. However, biocontrols of bacterial blight by those phages were unsuccessful, probably because changes in the phage population depend on that of the host population in nature. However, as a large population of the phages was detected from the irrigation water of severely infected fields prior to the disease appearance, a disease forecasting method was established and widely practiced in Japan from surveys of the phage population in paddy fields and channel water during the nursery and early paddy field stages where one to two bacterial isolates sensitive to phages prevalent in the survey area were used to enumerate the phage population.

Soil and environmental factors, such as nutrient availability, pH, temperature, moisture content and soil structure, affect virus–host interactions, and it is often the case in soil environments that unexpected, contradictory viral effects are observed for host control, although such results tend not to be published. The inefficiency of viruses often experienced in host control results, in part, from an inactive and consequently insensitive state of host cells in undisturbed soils. Under nutrient-limited conditions, phages are lytic when hosts are infected in their logarithmic phase, although they are pseudo-lysogenic for some time in the stationary phase.

Virus adsorption to Soils: A Key Factor in Virus Ecology in Soils

Virus adsorption to Soils

Adsorption of viruses to soils is a specific phenomenon profoundly influencing their infectivity to host organisms, and clay minerals play a central role in this phenomenon.

The adsorption follows either the Freundlich or Langmuir isotherm: the former under conditions of low surface coverage by viruses, and the latter under conditions of nearly saturated surface coverage. The amount of viral adsorption was found to be linearly related to the square root of time by Burge and Enkiri. In addition, Chattopadhyay and Puls suggested that viruses adsorb to colloidal clays using van der Waals forces. Phages were observed using TEM to adsorb to kaolinite particles by their tails.

The degree of adsorption of viruses to soils is generally more than 90% for all viruses. A variety of soil, environmental and viral factors affect the adsorption of viruses to soils. These parameters interact in viral adsorption on soils because viruses are, in general, negatively charged above the isoelectric point, whereas soils have negatively and positively charged sites as well as hydrophobic sites. The pH influences the charge on both viruses and soils, and ionic strength and its constituents determine the binding force between them. In contrast, organic matter (both soil-associated and dissolved) weakens the electrostatic binding of viruses to soils. The adsorption of the T2 phage of E. coli and Type 1 poliovirus on clay particles (kaolinite, montmorillonite and illite) was first demonstrated by Carlson et al. with an increase in viral adsorption associated with an increase in Na and Ca concentrations. The mechanism of viral adsorption to clays was via the clay–cation–virus bridge. Among cations, divalent cations were more effective as bridging agents than monovalent cations. And naturally occurring river clays (the Missouri River) showed very similar responses of viral adsorption to those of pure clay minerals.

Type of Clay Minerals and Type of Viruses

Clay minerals have both positively and negatively charged sites, and those sites contribute to viral adsorption to clay minerals. Schiffenbauer and Stotzky demonstrated that coliphages T1 and T7 had a greater affinity to montmorillonite than to kaolinite, and T7 showed a greater affinity to both clays than T1. T1 adsorbed primarily on positively charged sites of kaolinite and on both positively and negatively charged sites of montmorillonite. In contrast, T7 adsorbed more on negatively than on positively charged sites of both clays, which was reflected in an increase in its adsorption with an increase in the cation exchange capacity (CEC) of the clays. Reovirus type 3 had a similar adsorption tendency to T7 phage with more affinity to montmorillonite than to kaolinite, and a good correlation of adsorption to clays with CEC of the clays was observed because adsorption primarily occurred on negatively charged sites of the clays. A phage PBS1 of Bacillus subtilis behaved differently to reovirus Type 3 in its adsorption to clays, and there was no correlation between adsorption of PBS1 and CEC of montmorillonite and kaolinite. Reovirus Type 3 and coliphage T1 did not share common adsorption sites on montmorillonite and kaolinite, which suggests the importance of using more than one type of virus, especially in combination, to predict virus behavior (e.g. adsorption, loss of infectivity) in soils and sediments containing clay minerals. The clay minerals, kaolinite and montmorillonite, suspended in both distilled water and mineral medium (M-9), were better adsorbers for coliphage T1 or T7 than were bacteria, actinomycetes and yeasts. Except for the host bacteria, the microbial cells adsorbed few or no coliphages.

Carlson et al. first revealed the competition of egg and bovine albumin with viruses for adsorption sites on clay particles. Soils, especially soils rich in organic matter, are in general weak adsorbers compared with pure clay minerals. Lipson and Stotzky examined the effects of chymotrypsin, ovalbumin and lysozyme on the adsorption of a reovirus to kaolinite and montmorillonite. Chymotrypsin and ovalbumin suppressed viral adsorption to both types of clay minerals, whereas lysozyme only suppressed viral adsorption to montmorillonite. The suppression mechanism of these proteins on viral adsorption to clay minerals was the competition of proteins with the reovirus for sites on the minerals. Among the minerals studied, montmorillonite, glauconite and bituminous shale were the least effective, whereas the most effective adsorbers were magnetite sand and hematite, both of which are predominantly oxides of iron. Similar suppression effects of soil organic matter (mineral-associated and dissolved) on virus adsorption to soil were also observed for the somatic ssDNA φX174 phage and the unenveloped ssRNA MS-2 phage.

pH

Goyal and Gerba conducted a comparative adsorption study of a number of different types and strains of human enteroviruses and bacteriophages to nine different soils, and found that soils with a pH of less than 5 were generally good adsorbers. Loveland et al. studied the adsorption of an enteric phage PRD1 to quartz and ferric oxyhydroxide-coated quartz surfaces over a range of pH values and found irreversible adsorption of the virus to the surface below a critical pH (a pH approximately 2.5 to 3.5 pH units above the isoelectric pH point), but reversible adsorption above that critical pH.

Ionic Strength and its Constituents

Adsorption of viruses to low concentrations of clays was greater in estuarine water than in distilled water because the higher ionic strength of the estuarine water reduced the electrokinetic potential of both clay and virus particles. Thus, the addition of cations to distilled water enhanced viral adsorption, with divalent cations being more effective than monovalent cations. The sequence of the amount of adsorption to homoionic montmorillonite was $Al > Ca > Mg > Na > K$, whereas that to kaolinite was $Na > Al > Ca > Mg > K$. Soil adsorption was generally accelerated with the addition of Ca^{2+} and Mg^{2+} to the soil. Al^{3+} was the most effective cation in limiting poliovirus Type 1 penetrations through a soil column, probably owing to flocculation of the viruses, whereas organic compounds retarded the adsorption. Both natural humic material and sewage sludge organic matter increased the unsaturated flow of phage MS-2 in soil. As a counter anion, NO_3^-, SO_4^{-2} and $H_2PO_4^-$ anions were more effective than Cl^- in promoting the adsorption of poliovirus Type 1. However, virus adsorption on Al-oxide coated sand was suppressed in the presence of HPO_3^{2-} compared with HCO_3^-, and the effect of $H_2PO_4^-$

was more significant on phage φX174 than on phage MS-2. The presence of the divalent cations Ca^{2+} and Mg^{2+} suppressed viral adsorption on Al-oxide coated sand because the cations partially screened the negative charges on the phages, thereby decreasing the electrostatic attraction between the positively charged sand surface and the negatively charged phages. In addition, even small amounts of hydrophobic material in porous media can retard virus transport.

Viral adsorption to soil is highly strain dependent; less adsorption to soils was found by some echoviruses (Types 1, 2 and 29) and a simian rotavirus (SA1), whereas insignificant effects of specific surface area and pH were observed on poliovirus adsorption to soils. As coliphages MS2 and F2 adsorbed poorly to most soils, Bales et al. proposed to use them as an indicator for better estimating the maximum subsurface transport rate of colloidal contaminants through a porous formation. Although Burge and Enkiri found a fair correlation between the specific surface area of five soils and the adsorption rate of phage φX174 of E. coli, adsorption of reovirus Type 3 was not related to the surface area of kaolinite and montmorillonite.

Efficient buffer systems for extracting viruses from soils, therefore, are absolutely important to establish virus ecology in soils, although various organic and inorganic buffers have been examined for greater extraction of viruses from soils and sediments. Williamson et al. suggested the importance of multiple extraction and enumeration approaches before undertaking ecological studies of phages in a particular soil.

Movement of viruses in Soils (Epidemiological Concerns)

Epidemiological concerns in relation to virus ecology in soils are to trace the fate of pathogenic viruses discharged from wastewater treatment facilities and those present in aquifers and groundwater. The movement of viruses in soils is the opposite phenomenon of viral adsorption in soils, and is adversely related to viral adsorption to soils. Infiltration of viruses through soil columns, their soil surface run-off and their survival in soils were evaluated for various viruses, including coliphage, echoviruses, polioviruses, coxsackieviruses and rotavirus. Although soils are efficient adsorbers of viruses, viruses can migrate distantly in horizontal and vertical directions from several to dozens meters. Factors affecting the survival and migration of viruses in groundwater include soil texture, organic matter (humic acids), cations, adsorption property, pH, ionic strength, CEC, iron oxide level and permeability. In addition, virus type, microbial antagonism, rainfall, temperature, desiccation, flow rate and sunlight are also factors. However, rapid transport of the bacteriophage Salmonella typhimurium 28B was observed in a willow-cropped lysimeter experiment with clay soil compared with sandy soil because of bypass transport of the phage through macropore formation in the clay soil. Viruses penetrated deeper into the ground and moved longer distances than did coliform bacteria. Faster transport of F pilus-specific RNA coliphages than fecal coliforms was also observed in irrigated effluent through an alluvial aquifer near Christchurch, New Zealand. Jin et al. evaluated virus transport through sand columns

(9.2-cm diameter and 10.5 or 20-cm long) under saturated flow conditions using two bacteriophages, MS2 and φX174, and found that MS-2 was not sorbed to the Ottawa sand, whereas a significant amount of the applied φX174 was retained without inactivation in the columns, probably because of the difference in the isoelectric points of their surfaces. Similarly, Schijven et al. observed 3–10 fold higher transport rates of phage φX174 than of phages MS2 and PRD1, and this was attributed to the lower electrostatic repulsion of the less negatively charged φX174. Among five different spherical phages (MS2, PRD1, QB, φX174 and PM2) with different isoelectric points (pI 3.9, 4.2, 5.3, 6.6 and 7.3, respectively), the smaller phages (MS2, QB and φX174) with similar diameters exhibited maximum transport that correlated with their isoelectric points. In addition, when virus particles are more than 60 nm in diameter, their viral dimensions are believed to become an overriding factor.

In a survey of the rapid infiltration of wastewater containing a tracer virus, coliphage F2, indigenous enteroviruses and enteric indicator bacteria through unconsolidated silty sand and gravel, coliphage F2 concentration in a 18.3-m deep well began to increase within 48 h after application to the soil and stabilized after 72 h at a level of approximately 47% of the average applied concentration, irrespective of its concentration in the upper soil layers. In addition, indigenous enteroviruses and tracer F2 were sporadically detected in the groundwater at horizontal distances of 183 m from the application site.

Powelson et al. assessed soil-aquifer treatment of sewage effluent for removing bacteriophages MS2 (25 nm in diameter) and PRD1 (62 nm). PRD1 was removed more rapidly during percolation than MS2. Effluent type did not significantly affect viral removal rate for MS2, but the PRD1 removal rate was nearly threefold greater for secondary effluent than for tertiary effluent with virus removals at the 4.3-m depth ranging from 37 to 99.7%. The phage MS2 was adsorbed to clay minerals in the order of fine Na-montmorillonite > coarse Na-montmorillonite > Ca-montmorillonite and kaolinite, and the amounts of clay minerals that were transported through the columns followed the same order, which demonstrated that colloid-mediated transport was a dominant mechanism for virus transport through porous media. Kapuscinski and Mitchell emphasized the importance of viral association with colloidal and particulate materials in their transport in sewage coastal water management.

Environmental Factors affecting the Inactivation and Survival of Viruses in Soils:

A variety of soil and environmental factors affect viral inactivation and infectivity to host organisms, among which are temperature, pH, clay type and its content, organic matter content (nutrient status), heavy metals and acid pollutants, aerobicity (anaerobiosis), ionic strength (salinity), and moisture content.

Temperature

Bacteria are grouped into psychrophilic, mesophilic and thermophilic depending on their optimal growth/lethal temperature. Inactivation and survival of phages in soils are also temperature dependent, and the temperature dependency of the phages is in some cases not the same as that of their hosts.

Temperature is a major controlling factor for viral survival in soil. Yeager and O'Brien studied the effect of temperature (4, 22 and 37 °C) on the inactivation of poliovirus Type 1 in saturated soil and found that p.f.u. were recovered for up to 12 days at 37 °C, whereas at 4 °C p.f.u. were recovered from the soil for up to 180 days. Hurst et al. studied several viruses, coxsackieviruses (A9 and B3), echovirus Type 1, poliovirus Type 1, rotavirus SA11 and bacteriophages T2 and MS2, to elucidate the effects of temperature, soil moisture content, presence of aerobic microorganisms, degree of virus adsorption to soil, concentrations of resin-extractable P and exchangeable Al and soil pH on their survival, and temperature and virus adsorption to soil appeared to be the most important factors affecting their survival. Yates et al. revealed from multiple regression analysis that temperature was the only variable among numerous chemical and physical factors that correlated significantly with the decay rates of poliovirus, echovirus and coliphage in groundwater. In general, lower temperatures resulted in longer periods of survival, longer latent periods and reduced burst size of phages. It is probably because higher microbial and enzymatic activities, such as proteinase under higher soil temperature conditions, result in faster inactivation and decomposition of viruses. For example, phages of Geobacillus stearothermophilus (previously called Bacillus stearothermophilus) are ubiquitous in soil, and they were greatest in abundance at 45 °C when they were pre-incubated in soil with the host, whereas their proliferation was usually much less at 55 °C. Retention of plaque forming ability at 50 °C, but not at 70 °C, was also found for 24 phages of thermophilic Bacillus strains that were isolated from compost, soil, silage and rotting straw. Similarly, a phage infecting Bacillus circulans grew most at 45 °C compared with 30, 37 or 55 °C in a soil incubation experiment.

The inactivation of naturally occurring bacterial indicators of sludge and sewage contaminants (E. coli, fecal streptococci and spores of sulfite-reducing clostridia) and phages (somatic coliphages, RNA F-specific phages and phages infecting Bacteroides fragilis strain RYC 2056) was studied for the effect of thermal treatment on dewatered sludge at 80 °C and raw sewage at 60 °C, and those phages were shown to be significantly more resistant to thermal inactivation than bacterial indicators.

pH

Soil pH affects not only host growth, but also viral survival. Sykes et al. attempted to isolate the bacteriophages of five neutrophilic and three acidophilic streptomycetes from various soils with different pH levels, and no phage was detected in soils with a pH below 6.0, despite the presence of acidophilic streptomycetes in those soils. However, some phage isolates from neutral soils could lyse acidophilic streptomycetes at pH 5.5. Sykes and Williams found a loss in the infective potential of an actinophage to its host at pH 6.1 and 4.9 under the conditions of kaolin-adsorbed and free states, respectively. This was attributed to the accumulation of protons around the kaolin surfaces,

10-fold larger (1 pH unit), in comparison with the proton concentration in bulk soil water. Acidity might have multiple effects on adsorption, penetration and the length of the latent period in phage–host interactions.

Soil pH also influences viral survival indirectly. The adsorption of viruses to a solid surface is a common occurrence that results in a longer duration of survival. Viruses are hydrophilic colloidal materials, and their adsorption depends largely on the electrostatic properties of the soil surface to which the pH greatly contributes.

Clay Type and its Content

Poliovirus Type 1 and bacteriophages MS2 and PRD-1 added with sewage sludge to desert soils in Arizona, USA, survived for a longer period in a clay loam soil than in a sandy soil. Similar findings of better persistence and resultant suppression of nodule formation by phage Zag-1 of R. leguminosarum bv. trifolii in clay soil than in sandy soil were also reported from cowpea fields in Egypt. The rate of inactivation of phage φ11M15 of Staphylococcus aureus RN 450 in natural, autoclaved and filtrated lake water was greatly reduced in the presence of attapulgite and vermiculite and, to a lesser extent, of montmorillonite and kaolinite. Loss of reovirus infectivity in synthetic estuarine and distilled water containing montmorillonite or kaolinite was less than that in water without clays.

It has also been reported that viral association with colloidal and particulate materials prolonged survival in coastal waters. Although adsorption of poliovirus Type 1 to sandy loam soil protected the virus from inactivation, adsorption to sand had little effect on the rate of poliovirus inactivation. Lipson and Stotzky studied the mechanisms whereby particulates affect the specific infectivity of viruses. They used reovirus Type 3, kaolinite and L-929 mouse fibroblasts as the model system. Settling of the clay–virus complex on the L-929 cell monolayer increased the number of infectious virions reaching the cell surface. And the affinity of the reovirus to the cells was greater than to the clay, which was reflected in the difference in the energies of adsorption and binding between the virus and the cells or kaolinite. Thus, kaolinite enhanced the transfer of viral particles, in conjunction with diffusion and Brownian movement, to receptors of the reovirus on the host cell surface, primarily as a result of the more rapid settling of the kaolinite–virus complex on the cell surface. Clay minerals can protect phages against inactivation and loss of transducing ability by ultraviolet (UV) radiation, partly because they shield phages from UV light. Vettori et al. studied the effects of UV radiation on the survival of and transduction by phage PBS1 of B. subtilis under free or adsorbed states on montmorillonite and kaolinite. The titer and transduction frequency of clay-associated phages remained essentially constant between 1 and 10 min of UV radiation, whereas the titer of free phage decreased by approximately one order of magnitude after 5 min of UV radiation.

In addition, almost complete protection of E. coli from phage attack was provided by fine montmorillonite particles, although particles with an effective spherical diameter greater than 0.6 μm did not protect the bacterium from phage lysis.

Thus, clay minerals protect viruses against both biological and abiological inactivation, thereby enabling them to persist in soils for longer periods of time in the absence of their hosts.

Organic Matter Content (Nutrient Status)

Delisle and Levin tested plaque formation on a strain of Pseudomonas putrefaciens using psychrophilic phages that had been isolated from seawater and haddock meat and found that plaque formation was successful when the strain was grown on trypticase soy broth, but was unsuccessful when grown on trypticase soy broth with oxalate. This finding indicates that differences in substrate composition, for example, the kind of soils, influence the infectivity of phages to host bacteria, the susceptibility of the host bacteria, or a combination of the two in soils. One of the reasons of immunity of host cells to phage infection probably results from changes in the surface property of the host cells. For example, B. subtilis W23 was shown to lack wall teichoic acid when grown under conditions of phosphate limitation and did not adsorb phage SP50, the first step in phage infection. Lechevalier and Lechevalier analyzed some 600 strains of aerobic actinomycetes for their whole-cell sugar patterns and cell-wall compositions and classified them into 10 taxonomically different groups (Groups I to X). Polyvalent phages that were infectious to Streptomyces spp. infected hosts belonging to Group I genera, whereas all strains within Groups II, III, IV and VI (31 strains from 11 genera) were resistant to phages infectious to Group I genera.

Soil is an environment that generally provides poor substrates for microbial communities, and indigenous host cells may be in an insensitive state to viral infection in soil. Phages of Arthrobacter globiformis were rarely detected in soil, but they were continuously produced in nutrient-amended soil and the addition of host cells was not required for phage detection. Indigenous phages seem to be present in an inactivated state in soils, and the immunity of host cells to phages as a result of their physiological state may be phage specific or host specific. For example, Schrader et al. examined five phages specific for P. aerginosa and E. coli for their ability to multiply in host cells in stationary phase and found that the P. aeruginosa phage BLB was incapable of multiplication in a stationary phase host, but that significant multiplication of phage ACQ occurred in P. aeruginosa RM754 host cells that had been starved for periods of 24 h, several weeks or 5 years. Thus, starvation did not reduce bacterial hosts refuging from phage infection in the phage–host system of the latter.

Heavy Metals and Acid Pollutants

A 10 mmol L^{-1} concentration of Zn^{2+} did not significantly affect the survival of coliphages

T1, T7, P1 and φ80. However, the toxicity of zinc to those coliphages increased with the addition of high concentrations of NaCl (1 and 5 mol L^{-1}) because of greater toxicities of the anionic mixture of $ZnCl_3^-$ / $ZnCl_4^{2-}$ that was formed by NaCl addition. In contrast, mixtures of anionic $HgCl_3^-$ / $ZnCl_4^{2-}$ complexes were less toxic to phages (φ11M15 of S. aureus and P1 of E. coli) than were equivalent concentrations of cationic Hg^{2+}. The toxicity of 1 ppm Hg to φ11M15 was less in seawater than in lake water, indicating that the lower toxicity of Hg in seawater was a result of the formation of $HgCl_3^-$ / $HgCl_4^{2-}$ complexes in the seawater. Organic materials seem to alleviate the toxicity of heavy metals to viruses by forming organo-metal complexes. The toxicity of nickel to coliphage T1 was more serious at respective concentrations (5, 50 and 100×10^{-3} g Ni L^{-1}) in fresh water > estuarine water > sea water in this order. However, 100×10^{-3} g Ni L^{-1} was not toxic to T1 in nutrient broth, probably because of the protective effect of the organic compounds in the broth.

The effect of sulfur dioxide (SO_2) in acid rain on coliphage T1 was evaluated in a test tube by changing pH and HSO_3^- and SO_3^{2-} concentrations. HSO_2^- was always more inhibitory than SO_3^{2-}, and the inhibitory effects of HSO_3^- and SO_3^{2-} increased with a decrease in pH. No study has been conducted on the mechanism of toxicity of heavy metals and acid pollutants to viruses.

Aerobicity (Anaerobiosis)

Replication under anaerobic conditions resulted in longer latent periods of phages nt-1 and nt-6 of Vibrio natriegens. However, the effect of anaerobiosis on burst size was different between these phages: phage nt-6 had a reduced burst size, whereas nt-1 had an increased burst size. Therefore, the composition of phages in lowland areas, including rice fields, may vary according to reclamation and water management.

Ionic Strength (Salinity)

Two phages (nt-1 and nt-6) of the marine bacterium Beneckea natriegens (Vibrio natriegens) were examined to determine the effect of ionic strength on their stability and replication. They were more resistant to low NaCl levels than their host bacterium, which appeared to confine their growth to marine waters because of the optimal higher salinity of B. natriegens survival. Moebus summarized the effects of ionic strength of Ca^{2+}, Mg^{2+} and Na^+ on adsorption on host cells, burst size, latent period and the survival of phages in seawater. Virulence of the phage PK-101 of Pseudomonas solanacearum K-101 increased with the addition of $CaCl_2 - MgCl_2$ (0.15 mol L^{-1}). Adsorption of viruses to clay particles is greater under higher ionic strengths because higher ionic strengths reduce the electrokinetic potential of both clay and virus particles. And viral association with colloidal and particulate materials prolongs the survival and infectivity of viruses. Thus, ionic strength is an important environmental factor coupling with clay particles for the survival of viruses.

Moisture Content

Soil drying results in profound effects on viral inactivation. There appears to be a critical point of soil moisture below which virucidal effects of evaporation are amplified. For example, poliovirus Type 1 was inactivated at essentially the same rate between 18 and 2.9% moisture content, but the rates of viral inactivation increased markedly between 1.2 and 0.6% moisture for a sandy loam soil. Viral inactivation by soil drying resulted from the virucidal effects of the evaporative process per se as well as from the effects of low soil moisture levels, and evaporation was estimated to be the primary factor responsible for viral inactivation in drying soil for poliovirus Type 1. The irreversible binding of viruses that may occur as a result of drying in soil environments is another factor affecting viral inactivation. Yeager and O'Brien used radioactively RNA-labeled and capsid-labeled poliovirus Type 1 and found that both the RNA and capsid portions of the virions were recovered from moist soils, but that only the RNA label was recovered efficiently from the dried soils. Inactivation of poliovirus also occurs in moist soils because of the damage of RNA before its release from the capsid. In contrast, the dissociation of virions into intact RNA and isoelectrically altered capsids was noted in dried soils. Inactivation of poliovirus Type 1 strain CHAT by cleavage of vrial proteins followed by nicking of encapsulated RNA was also observed in digested sludge.

To assess sanitary risks by pathogenic, enteric viruses present in the sludge, laboratory studies were conducted to measure the inactivation rate of poliovirus Type 1 and bacteriophages MS2 of E. coli and PRD-1 of Salmonella typhimurium in two sludge-amended desert agricultural soils (sandy loam and clay loam soils). As the temperature increased from 15 to 40 °C under constant moisture (30% content, approximately -0.05×10^5 Pa for both soils) and temperature conditions, the inactivation rate increased significantly with the increase in temperature for poliovirus and MS2, whereas a significant increase in the rate was observed only at 40 °C for PRD-1. For all three viruses, drying to less than 5% soil moisture resulted in a more rapid loss of infectivity than at a constant moisture. Evaporation completely inactivated all three viruses within 7 days at 15 °C, within 3 days at 27 °C and within 2 days at 40 °C, regardless of soil type. Williamson et al. measured viral-like particles (VLPs) in six Delaware soils, USA, using epifluorescence microscopy and found a significantly positive correlation between VLP abundance and soil water content.

In addition, as Williamson et al. found a significant difference in VLP abundance between agricultural soils and forest soils, land-use form and land management practices might be important underlying factors controlling viral abundance in soil environments.

Host Specificity of Phages in Soil

Host specificity of rhizobiphages may match the infection range of Rhizobium spp. Eleven phages of Rhizobium loti, R. leguminosarum bv. trifolii, Rhizobium galegae and Rhizobium meliloti were isolated from fields where legumes grew and they were used for phagotyping of nodule bacteria with unknown taxonomic position from milk vetch (Astragalus), liquorice (Glycirrhiza), rest-harrow (Ononis), French honeysuckle (Hedysarum) and crown vetch. Phages of R. meliloti, R. leguminosarum and R. galegae lysed only bacterial strains of homologous species. However, in a follow-up study, Novikova et al. did not find a relationship between the infectivity of phages to various Rhizobium spp. isolated from R. loti, R. meliloti, R. leguminosarum and R. galegae root nodules and the nodule-forming capacity of those Rhizobium strains among legume plants. Hashem et al. proposed a rapid method for identifying and characterizing Sinorhizobium fredii strains from their susceptibility to various phages. Host specificity of phages to X. oryzae pv. oryzae (pathogen of bacterial blight of rice plant) is different according to the strains of phages and isolates of the bacterium. And the bacterium was classified according to the lysotype to various phage strains. Although the reason was not clear, Kankila and Lindström found an inverse relationship between the host range of phages of R. leguminosarum bv. trifolii and the level of susceptibility of phage DNA to restriction enzymes. Host range of phages is, in general, within the genus level.

Horizontal Gene Transfer by Viruses

Transduction is a mechanism for gene transfer mediated by viruses among bacteria. There are two mechanisms of transduction: generalized transduction and specialized transduction. In generalized transduction, lytic phages package DNA from the infected bacterium, whose chromosome has been cleaved into small segments by nucleases, into the phage capsid. As any single or group of bacterial genes may be incorporated in this transduction, it is designated as generalized transduction. The transducing phages carrying bacterial DNA inject their DNA, including the bacterial DNA fragment, into new bacterial host cells. Specialized transduction is mediated by a lysogenic phage, where the phage genome is integrated into the bacterial chromosome, usually at a specific site, and replicates faithfully with the bacterial chromosome. Upon induction of the lytic cycle of phage replication, imprecise excision of the phage genome results in the packaging of adjoining bacterial genes on the chromosome into the virion with the prophage DNA. Thus, as this transduction transfers only genes that are close to the site of phage integration, it is termed specialized transduction.

Generalized transduction of E. coli K-12 by P1 coliphage was demonstrated using auxotrophic recipient cells (thr-leu-thi-rpsL) and transducing phage lysates from transposon 10 (Tn10) in sandy and silty clay loam soils. The frequency of transduction in soil was approximately $10-6$. The specialized transduction of bacterial resistance genes for chloramphenicol and mercury into E. coli was demonstrated when the transducing coliphage P1 was added to non-sterile soil as either lysates or E. coli lysogens. Ripp and Miller examined the effects of mackaloid clay and particulate matter in freshwater from Lake Carl Blackwell water, Oklahoma, USA, on the frequency of transduction among Psuedomonas aeruginosa mediated by temperate phage F116, and observed an increased frequency in generalized transduction by 100-fold, because aggregations of phages and bacterial cells are stimulated by the presence of these suspended particulates.

Lysogeny is common and occurs frequently in bacterium–phage relationships, and up to 40% of isolates of P. aeruginosa from natural ecosystems were found to contain DNA sequences homologous to phage genomes, evaluated using radiolabeled bacteriophage DNA probes. Lysogenization is a mechanism of gene transfer in soil among closely related bacterial communities. Herron and Wellington observed that the temperate phage KC301 incorporated into Streptomyces lividans TK23 (S. lividans TK23 [KC301]) was recovered from S. lividans TK24 as S. lividans TK24(KC301) in soil. Transfer of KC301 from TK24 (KC301) to a Streptomyces violaceolatus strain was also recognized in natural soil amended with nutrients. In general, the lysogenized organisms are considered to be less adapted to the soil environment in comparison with the original prophage-free organisms because of DNA reproduction of inserted prophage sites, although Ashelford et al. did not observe any difference in the survival of Serratia liquefaciens CP6RS and its lysogen (CP6RS-ly-Φ1), which had been inoculated to sugarbeet seeds during a 194-day growth period.

Enumeration of Viruses in Soils

For accurate evaluation of viral abundance, reliable methods of extraction of viruses from soils, separation of viruses from bacterial communities, storage of extracted viruses, and their enumeration are required.

Extraction of Viruses from Soils

As the majority of viruses are adsorbed to the solid phases of soils, solutions of various composition have been devised for better extraction of viruses from soils and sediments; for example, nutrient broth, nutrient broth containing egg albumen, soil extract, beef extract, beef extract with NaCl, beef extract containing borate, glycine buffer, sodium pyrophosphate, potassium citrate, $AlCl_3$ (pH 3.5), and mineral buffer. In general, beef extract (6–10%) has most widely been used for the extraction of viruses from soils.

Separation of Viruses from Bacterial Communities

Filtration through membranes of 0.22–0.45 μm pore size is most commonly used for the separation of viruses from bacterial communities. The procedure, in general, results in a significant loss of viruses in extracts, with a two-third reduction in some cases. Moreover, the selection of the filter material is important because of viral adsorption on filter membranes. It should be noted that separation of viruses from bacterial communities using such filtration techniques results in disregarding viruses in host cells.

Storage of Extracted Viruses

A 2–2.5% final concentration of glutaraldehyde is generally used for the preservation of water samples and viral extracts in addition with 2% final concentration of formaldehyde. However, as formaldehyde and glutaraldehyde are known to induce significant reductions in virus counts after 24 h of storage of marine sediments, immediate measurement of viral abundance in water samples and extracts is recommended. Deep freezing at −20 °C or colder is recommended for viral storage.

Enumeration

As the infectivity of viruses is specific with respect to hosts, enumeration of viral abundance by plaque formation is used only for detecting and isolating specific viruses. Because of the very strict host specificity of viruses, precondensation of samples and/or enrichment culture are generally necessary. The double agar layer method developed by Adams is the standard method for plaque counting. Enumeration of viral abundance by TEM has been commonly used since the first application of TEM to viral enumeration in seawater by Torrella and Morita, who counted phage particles collected on a 0.2-μm Nuclephore filter. They observed different morphologies of phages as well as phage–bacterium interactions. Specimen preparation for the quantitative collection of viral and bacterial communities in water samples and extracts on microscope grids was developed by Nomizu and Mizuike. Although TEM is the established method for viral enumeration in water samples and viral extracts and has been commonly used in the past and at present, epifluorescence microscopy has been introduced for viral enumeration in the environment because of the reasonable price of equipment, easy sample preparation and a similar degree of accuracy to the TEM method since the method was applied to marine waters by Hara et al.. Yo-Pro, YoYo-1, PoPo-1, DAPI and SYBR Green I have been used as fluorescent dyes for the estimation of viral abundance in marine and sediment samples. SYBR Green I enabled the rapid and accurate determination of viral abundance collected on 0.02-μm pore-size filters after staining for 15 min. Viral counts by SYBR Green I were, on average, 1.28-fold higher than those made by TEM. These researchers attributed the underestimate of viral abundance by TEM to a loss when uranyl acetate is wicked away from the grids and viruses may be obscured by other larger, darkly stained particles on the grids.

Abundance of viruses and their roles in biogeochemical nutrient cycles in aquatic environments. The importance of viruses in biogeochemical nutrient cycles and as genetic reservoirs in marine and freshwater environments results from their large abundance and great diversity. However, comparable information on viral abundance in soil environments has only been collected by Williamson et al. and Nakayama et al. Then, the significant role of viruses in biogeochemical nutrient cycles in aquatic environments is summarized. The importance of viruses in this matter results from the large abundance of viruses, the short generation time of host organisms, and the significant proportion of virus-mediated mortality in aquatic environments.

Abundance

Viral abundance increases with the productivity of the system, and it is generally lowest in the deep sea ($10^4 - 10^5$ VLP mL^{-1}), intermediate in offshore surface waters ($10^5 - 10^6$ VLP mL^{-1}) and highest in coastal environments ($10^6 - 10^7$ VLP mL^{-1}). Viral abundance in estuaries or productive lakes can be as high as 108 VLP mL−1, with the highest viral number of 9.6 × 108 VLP mL^{-1} recorded in a cyanobacterial mat. Viral abundance tends to be higher in freshwater than in marine systems.

Numerous studies have examined temporal and spatial variations in total viral abundance. Sunlight and annual phytoplankton blooms are characteristic factors for viral abundance in aquatic environments. Viral abundance is larger where suspended particles and substrates are larger. Thus, factors controlling viral inactivation and abundance are different from those in the soil environment, where temperature, pH, clay particles, ionic strength (salinity) and moisture content are important factors. In marine environments, temperature and salinity are stable and seem not to affect viral communities directly, but they appear to be correlated with microbial production (e.g. phytoplankton blooms) and substrate concentration, such as the decrease in viral abundance along the transect from freshwater, estuarine to open ocean. The pH is generally stable in aquatic environments. Moreover, pelagic environments are generally free from clay particles in the strict sense.

In general, viral-to-bacterial ratios (VBRs) are between 3 and 25, and are higher in more nutrient-rich, productive environments, which is one of the reasons for viruses being the greatest genomic reservoirs. Both positive and negative relationships between VBR and bacterial abundance have been observed in oligotrophic and eutrophic seawater. Although positive relationships were noted in both oligotrophic waters in the North Pacific and eutrophic waters in the northern Adriatic Sea, the most commonly observed relationships are inverse between them in various seawaters. In the floodwaters of a Japanese paddy field, the relationship was also significantly inverse between them in various seawaters.

Studies on viral abundance in sediments are surprisingly few in comparison with the voluminous studies in water columns in aquatic environments. Viral and bacterial abundances in sediments are of the order of $10^7 - 10^9$ VLP mL^{-1} and $10^6 - 10^8$ cells mL^{-1} of sediment, respectively, with VBRs fluctuating from 0.1 to 100. As the measurements were conducted with either pore water or 10 mmol L−1 pyrophosphate buffer in these studies, wide variations in VBRs seem to be derived from the specificities of the study locations.

The VLP and bacterial abundances in the floodwater of a Japanese paddy field ranged from 5.6×10^6 to 1.2×10^9 VLP mL^{-1} and from 9.2×10^5 to 4.3×10^8 cells mL^{-1} during the period of rice cultivation from June to September, with mean abundances of 1.5×10^8 VLP mL^{-1} and 5.1×10^7 cells mL^{-1}, respectively, indicating that the floodwater was rich in viral and bacterial abundances among various aquatic environments, and abundances increased with an increase in the turbidity of the floodwater. Thus, the magnitude of seasonal variation was more than 50-fold for VLP abundance and 100-fold for bacterial abundance.

Williamson et al. extracted 10^9 VLP g^{-1} of soil with potassium citrate buffer from six Delaware soils, USA, under different vegetation (two agricultural soils and four forest soils). The abundance of VLPs was highest in the wetland forest soils ($2.9 - 4.2 \times 10^9$ VLP g^{-1}) followed by upland forest soils ($1.3 - 1.5 \times 10^9$ VLP g^{-1}) and agricultural soils ($0.87 - 1.1 \times 10^9$ VLP g^{-1}). The VBRs were extremely high in the agricultural soils (approximately 3000), whereas the VBRs in the forested soils were approximately 10. The extremely high VBRs in the agricultural soils were attributed to less efficient extraction of bacteria from agricultural soils. This was the greatest record of VLP abundance in soil ever published. Their results suggested seemingly conflicting strategies used by viruses to exist in soils: the large VBRs indicate that the virulent phages are dominant in those soils (the environment for r-strategists) even if the VBRs were approximately 10 in those soils. In contrast, the viral abundance per se indicates that lysogeny is a common lifestyle of viruses in soils (the environment for K-strategists) because most soils commonly contain bacteria in a range from 10^8 to 10^{11} cells g^{-1}. More information on viral abundance in soils with more reliable extraction methods is a prerequisite for evaluating the roles of viruses in biogeochemical nutrient cycles and genomic diversity in soils.

Microbial loop in biogeochemical nutrient cycles in aquatic environments. Many studies indicate the ecological importance of viruses in primary production and microbial food webs in marine environments. At present, it is generally estimated that viruses are responsible for approximately 10–50% of the total bacterial mortality in marine surface waters, and 50–100% in environments that are hostile to protists, such as low-oxygen lake waters. The viral effect is probably larger in eutrophic waters than in oligotrophic waters. Figure shows the pelagic food chain model and virus-mediated carbon flow constructed by Weinbauer, where 6–26% of the carbon fixed by primary producers is estimated to enter into the dissolved organic carbon (DOC) pool via virus-induced lysis at different trophic levels. The situation is similar in freshwater environments, and Fischer and Velimirov estimated that the viral control of bacterial production ranges from 56 to 63%, with occasionally up to 1.6% h^{-1} of the bacterial standing stock being removed from the water column, while grazing of heterotrophic nanoflagellates accounted, on average, for only 5% of the bacterial mortality in a eutrophic oxbow lake of the Danube River.

Much of the DOC derived from viral lysis of bacteria is not transferred to higher trophic levels, but is recycled through the microbial community, which is termed the "microbial (bacterium–phage–DOC) loop". The total production of heterotrophs can greatly exceed primary production because of the recycling mediated by phages in the marine environment. Thus, repeated cycling of organic materials in the microbial loop causes the bacteria to be efficient sinks for C and in the regeneration of N and P in environments where viruses are important agents of microbial mortality.

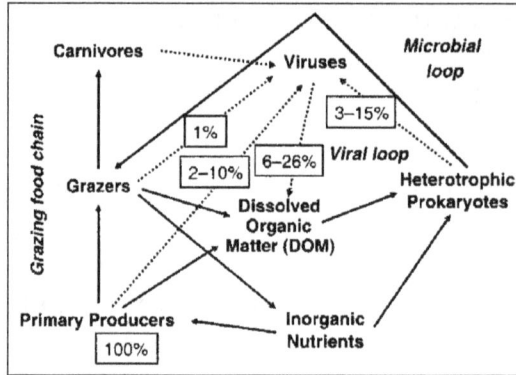

Pelagic food chain model and virus-mediated carbon flow. The dotted lines show virus-mediated pathways. All values are based on the flux of carbon fixed by primary producers (100%).

The importance of viruses in the microbial loop in biogeochemical nutrient cycles in marine environments is a result of the large amount of microbial biomass and their fast mortality rate by viral lysis. According to the summaries by Weinbauer (2004), as phage-infected bacterial cells ranged from 1.4 to 26% in oceanic waters and from not detectable to 43% in coastal/shelf waters in bacterial communities, phage-mediated mortality of bacteria was 25% in oceanic waters and 58% in coastal waters. As viral turnover time ranged from 0.1 to 25 days, roughly 10–20% of bacterial members are estimated to lyse through phage infection on a daily basis in marine environments.

The significant role of viruses in biogeochemical nutrient cycles is only as a driver of the microbial loop. The total biovolume of VLPs ranged from 0.027 to 2.9 cm^3 m^{-3} with a median volume of 0.31 cm^3 m^{-3} in the floodwater of a Japanese paddy field, where the virus-to-bacterium biovolume ratios were estimated to be less than 0.002 because the ratio (0.002) referred to the case of bacterial communities being coccal with a diameter of 0.45 µm (the pore size diameter of the Nuclepore filter used for bacterial enumeration by epifluorescence microscopy). As viruses are functionally active only inside host cells, their life cycles synchronize to those of the host organisms. In addition, not all the host cells are burst by viruses synchronously. Therefore, the contribution of biovolume of virus communities per se is negligible in nutrient cycles in the floodwater in comparison with that of bacterial communities, which is in contrast to the important contribution of microbial biomass in nutrient supply to crops in soils.

Lytic viruses may contribute more to the activation of the microbial loop in the biogeochemical nutrient cycles than lysogenic viruses. And as a greater incidence of lysogenic bacteria was isolated from oligotrophic marine environments compared with coastal or estuarine environments, the contribution of viruses to biogeochemical nutrient cycles seems to be less significant in environments such as the pelagic ocean. The situation is not so simple in other environments, and Jiang and Paul showed that the percentage of lysogens that was estimated from prophage induction by mitomycin C, UV radiation, sunlight, temperature and pressure was highest in estuarine waters among estuarine, coastal and oligotrophic offshore environments. We have to remember that the existing lysogens are the remnants after factors such as detrimental temperature, pH, ionic strength or mutagens have already conducted a significant proportion of the host communities in a sample to prophage induction. In summary, the contribution of lysogenic viruses in biogeochemical nutrient cycles compared with that of lytic viruses might depend on the degree of induction of lysogens by natural stresses in respective environments in time and space.

Lysogeny seems to be chosen in the evolutionary process by viruses as a survival strategy at low host abundance and activity (K-strategic). Wilcox and Fuhrman observed in coastal surface seawater that lytic infection occurred only when the product of viral and bacterial numbers equaled 10^{12} mL^{-1} or more. However, because of the specificity of virus–host interactions, infection is also dependent on bacterial species composition and diversity. This value may be different in soil environments because the threshold value was determined mainly by the frequency of viral encounters with bacterial hosts in seawater. Higher values may be expected from the adsorption of most viruses on soil particles, resulting in the restriction in the free movement of viruses to encounter their hosts, and lower values from the accumulation of viruses and hosts on soil particles.

The microbial world in soils is very different from that in aquatic environments in terms of composition and turnover time of microorganisms. Fungal biomass is often larger than that of bacteria in soils. And the turnover time of microbial communities is, on average, 1 year or longer in undisturbed soils. No attention has been paid to the viral mortality of soil microorganisms that may contribute to the determination of turnover time of soil microorganisms. Moreover, as higher plants are the primary producer and the microbial loop, if any, is driven predominantly by heterotrophs growing on plant residues and humic substances in soils, the significance of the microbial loop in biogeochemical nutrient cycles in soil and the roles that viruses may play in it are different from those in aquatic environments.

Diversity of Viruses and their Importance as Genomic Reservoirs in Aquatic Environments. Until recently, direct TEM examination of aquatic viruses was the most common method of recording viral diversity in aquatic environments. Although morphological data give only a limited view of viral diversity, these data have been cited as evidence that bacteriophages comprise the majority of viruses within viral communities. Many observations of temporal and spatial changes in the frequency distribution of viral capsid sizes have been recorded for viral populations. Now evaluation of viral diversity and exploration for novel genes in aquatic environments from genomic analyses open a new view of aquatic virus communities, the studies of which are the central topics of "the third age of virus ecology".

Size Distribution of Viruses

Viruses in aquatic environments are generally dominated by the 30–60-nm capsid size class. The dominant viruses in the flood water of paddy fields also fell into a similar size class Size distribution changes temporally and spatially from coastal to open ocean sites, with depth, and over a 1-month sampling period during spring diatom blooms. A difference in viral size class distribution was also noted among water samples collected from 22 Canadian lakes.

Not every study has documented the change in the distribution of capsid size classes with a change in time or the location of the water samples collected. Mathias et al. noted no change in the frequency of four classes of capsid diameters among the virioplankton in river water samples collected over a 2-year period. Nakayama et al. also did not find a change in capsid size distribution in the floodwater of a Japanese paddy field under different fertilizer treatments (no fertilization; chemical N, P and K fertilizers; chemical N, P, K and Ca fertilizers; and compost with chemical N, P, K and Ca fertilizers) during the entire period of field flooding from transplanting to harvesting.

Weinbauer and Peduzzi showed for Adriatic Sea water samples that the 30–60-nm capsid size class comprised 74 and 100% of intracellular viruses in rods and spirilla, respectively, whereas cocci more often contained larger viruses (60–110 nm) than the smaller 30–60 nm viruses (65%vs 35%) in the cells. This finding is indicative of the predominance of the 30–60-mn capsid size class of viruses in the marine environment because bacterial communities are predominated by rod-shaped members.

The predominance of viruses within the 30–60-nm size class is attributed to the predominance of phages in viral communities and the selective predation of larger viruses by heterotrophic nanoflagellates. The average capsid size for viruses of eukaryotic algae is reported to be 152 nm, although 28% of them are less than 60 nm. In general, phages are smaller, ranging from 34 to 160 nm with a peak sharply at 60 nm based on 251 phages. Ackermann and DuBow also found that phages with isometric capsids have a clear preference for the 55–65-nm range from approximately 180 species descriptions of classified phages of various bacteria. Viruses with different capsid sizes were ingested at different rates, with the smallest virus being ingested at the slowest rate. And these researchers calculated that when there are 106 bacteria mL^{-1} and 10^7–10^8 viruses mL^{-1}, viruses may represent 0.2–9% of the carbon, 0.3–4% of the nitrogen and 0.6–28% of the phosphorus that the flagellates obtain from the ingestion of bacteria.

The clonal diversity of viral communities is evaluated on genome size using pulsed-field gel electrophoresis (PFGE), although the resolution of PFGE is not very high and the number of bands on the agarose gel is a conservative estimation of the number of viral species. Wommack et al. used PFGE to monitor the population dynamics of Chesapeake Bay virioplankton for an annual cycle. The PFGE analysis detected several distinct bands ranging from 50 to 300 kb, and statistical analyses elucidated changes in virioplankton community structure in relation to sampling time, geographical location and the extent of water column stratification. And the hybridization of PFGE-separated samples with DNA probes specific to single viral strains and a group of viruses with similar genome sizes demonstrated that the abundances of specific viruses changed in time and space in Chesapeake Bay.

Morphology of Viruses

Although most studies on the morphology of free viruses in aquatic environments did not present quantitative data on the proportions of tailed versus non-tailed forms, most free viruses in aquatic environments appear to be predominantly tailed forms. The exception was the study by Tapper and Hicks, in which they recorded the presence or absence of tails as well as capsid size distribution in water samples from Lake Superior, Canada. On average, 70% of viruses were tailed and the proportion of tailed viruses increased with an increase in the capsid size from 65, 74 to 100% for those with < 30-nm, 30–60-nm and > 60-nm capsid diameters, respectively.

In six Delaware soils studied by Williamson et al., most soil viruses were phages belonging to Caudovirales. Among the tailed phages, podophages and siphophages were more dominant than myophages in these soils. In addition, a much higher frequency of filamentous and elongated capsids was found in these soils. In particular, approximately 10% of the viral community in a silt loam soil consisted of elongated capsid phages. The predominance of tailed viruses in aquatic and soil environments is natural because 96% of phages were tailed among over 5100 phages examined by TEM. It is important to note that non-tailed forms may include free viruses that have lost their tails during the preparation for TEM observation.

Genomic Diversity in Aquatic Viral Communities

Although Siphoviridae or phages with long, non-contractile tails comprised 61% of tailed phages in the examinations using TEM since 1959, a predominance of Siphoviridae among tailed viruses is not necessarily the case in marine environments. Breitbart et al. examined a viral-community DNA library created from the Mission Bay sediment, California, and from two marine waters at Mission Bay and Scripps Pier, California, using linker-amplified shotgun cloning. They found a predominance of T7-like podophages, λ-like siphophages and T4-like myophages in two marine waters and λ-like siphophages, T7-like podophages and T4-like myophages in the sediment, in this order, among clones significantly similar to previously reported phage sequences, although more than 65% of the sequences were not affiliated with any known sequences. In contrast, a majority of phage isolates from marine environments were generally myophages and siphophages.

Until recently, it was commonly understood among viral taxonomists that there is no single gene that is common to all virus genomes and that a total diversity of uncultured virus communities cannot be evaluated using approaches analogous to 16S ribosomal DNA as for bacteria. However, the genome of a lytic phage P60 of marine Synechococcus

WH7803 contained approximately 47,872 bp with 80 potential open reading frames that were mostly similar to the genes found in lytic podophages (T7, φYeO3-12 and SIO1). In addition, according to the 109 bacteriophage genomic databases available in 2001, lysogenic phage was the dominant form among the known myophage and siphophage genomes and integrase genes were found in all the lysogenic myophage and siphophage genomes, whereas eight of nine podophage genomes contained the DNA polymerase, primase and helicase genes. It was also noted that three Prochlorococcus phages (a podophage and two myophages) were quite similar to T7-like (P-SSP7) and T4-like (P-SSM2 and P-SSM4) phages containing 15 of 26 core T7-like genes and 43 and 42 of 75 core T4-like genes, respectively. These findings indicate that comparable genomic information is preserved among viral subsets and that genomic information may be an effective tool for the phylogenetic classification of viruses and the phylogenetic evaluation of viral diversity in the environment. Metagenomic analyses are elucidating the remarkable diversity of environmental viral communities, and Breitbart and Rohwer estimated that there were possibly 5000 viral genotypes in 200 L of seawater and one million different viral genotypes in 1 kg of marine sediment.

Furthermore, some functional gene sequences in the host bacterial genomes were found to be highly conserved in viral genomes, examples of which are the DNA polymerase gene pol, the photosynthesis genes psbA and hliP, the aldolase family gene talC, and the phosphate-inducible genes phoH and pstS. These findings substantiate the possibility of phylogenetic classification of viral subsets in the environment by comparing these genomic sequences. Rohwer and Edwards presented the "Phage Proteomic Tree" based on the overall similarity of 105 completely sequenced phage genomes. The resulting taxonomy was compatible with the system of International Committee on the Taxonomy of Viruses (ICTV), indicating that phage taxonomy has entered the post-genomic era. The presence of common functional genes related to fundamental physiological processes, energy-acquiring processes and gene expression processes across virus, prokaryote and eukaryote kingdoms strongly indicates horizontal gene transfer among the kingdoms, which may contribute to an increase in the biodiversity of respective genes via transduction. In addition, phage-encoded virulence factors are found in a variety of phages, members of Myoviridae, Siphoviridae, Podoviridae and Inoviridae, with some phages having characteristics of more than one family. A phage–bacterium interaction is, therefore, not a simple parasite–host interaction, but an instance of coevolution of phages and prokaryotic cells.

The following topics provide brief summaries of the viral diversity in aquatic environments estimated from the structural genes (g20, g23) of T4-like phages, the polymerase gene (pol) of T7-like phages, and the photosynthesis gene (psbA) of marine cyanophages, as well as evidence of horizontal gene transfer among the living world.

Structure Genes of T4-like Phages

A marine cyanomyophage S-PM2 has a genome homology to coliphage T4 at a 10-kb region as a contiguous block from gene g18 to g23. In T4, g18 codes for the tail sheath, g19 for the tail tube, g20 for the head potal proteins, g21 for the prohead core protein, g22 for a scaffolding protein, and g23 for the major capsid protein. The T4-like phage family was further classified into subgroups of the T-evens, PseudoT-evens, SchizoT-evens and ExoT-evens, with increasing divergence from T4 based on the sequence comparison of g18, g19 and g23 genes.

The polymerase chain reaction (PCR) primers CPS1 and CPS2, which specifically amplify a 165-bp region from the majority of cyanomyophages, were constructed first by Fuller et al., although priming efficiency of the primers exhibit phage-to-phage variability. The region has significant similarity to g20 of coliphage T4. Primer pairs of CPS1 and CPS8 constructed by Zhong et al. cover more nucleotide sequences of g20 and elucidated the presence of nine phylogenetic groups among 114 totally different g20 homologs in six natural virus concentrates from estuarine and oligotrophic offshore environments. Only three groups/clusters contained known cyanophage isolates, and the identities of the other six clusters remain unknown.

The analysis of cyanophage communities by the g20 gene is also useful in the natural freshwater environment. Dorigo et al. studied cyanophage communities over time in Lake Bourget, France, using CPS1–CPS8 primers and found 35 distinct cyanomyophage g20 genotypes among 47 sequences analyzed. Phylogenetic analyses showed that these sequences fell into seven genetically distinct operational taxonomic units (OTUs). Some of these freshwater cyanophage sequences were genetically more closely related to marine cyanophage sequences than to other freshwater sequences. Similar findings of a wide distribution of closely related hosts and/or horizontal gene exchange among phage communities from very different environments were also obtained by analyzing g20 genes in various water samples from the Gulf of Mexico, the Arctic, the Southern, Northeast and Southeast Pacific Oceans, an Arctic cyanobacterial mat, a catfish production pond, lakes in Canada and Germany, and a depth of approximately 3246 m in the Chuckchi Sea. These researchers also found four novel phylogenetic groups of g20 genes, among which two were only found in freshwater.

In addition, Mühling et al. examined the control of Synechococcus genotypes by phages in the oligotrophic Gulf of Aquba, Red Sea, over an annual cycle from denaturing gradient gel electrophoresis (DGGE) patterns of a 118-bp g20 gene fragment of cyanophages and Restriction Fragment Length Polymorphism (RFLP) patterns of a 403-bp rpoC1 gene fragment of Synechococcus spp., in which they found that both the abundance and genetic diversity of cyanophage communities covaried with those of the Synechococcus communities. Wilhelm et al. observed the pervasive distribution of cyanophages in Lake Erie, USA, which can infect the marine Synechococcus sp. strain WH7803, and analyses of g20 indicate that these phages are related to marine cyanophages, but in some cases form a unique clade, leaving questions with regard to the natural hosts of these phages.

Filée et al. compared the phylogenies constructed from g23 gene segment sequences with those obtained from the T4-type phage genomes for 16 completely sequenced T4-type phages and found a very good similarity between the phylogenies. Therefore, they designed degenerate primers targeting the g23 gene of phage T4 (MZIA1 bis, MZIA6) and applied them to elucidate T4-like bacteriophage communities in diverse marine environments (fjords and bays of British Columbia, the eastern Gulf of Mexico, and the western Arctic Ocean). Although some of the sequences of the PCR products were closely related to well-studied subgroups of the T4-like phages, such as the T-evens, SchizoT-evens, PseudoT-evens and ExoT-evens, the majority belonged to five previously uncharacterized subgroups. Thus, the g23 gene has more appropriate regions than the g20 gene for the phylogenetic evaluation of T4-type phage communities in the environment.

However, g23 genes may not be specific for T4-type phages. Jenkins and Hayes recently compared amino-acid sequences of g23 fragments among 17 cyanophage isolates of the heterocystous, filamentous cyanobacterium Nodularia spumigena using-specific primer set (CAP 1 and CAP 2). Although they were diverse in terms of their morphology and host range and belonged to two families of Myoviridae and Siphoviridae in Caudovirales, the encoded protein was 99% identical to T4 g23 homologues across all cyanophages compared. This fact indicates that g23 genes may be shared among Myoviridae and Siphoviridae members and that amino-acid sequences of the major capsid protein (g23) are important in phage–host interactions, irrespective of the phylogenetic positions of the phage and the host.

Primer pairs (MZIA1 bis, MZIA6) were also used by Jia et al. to compare DNA extracts of surface soil and rice straw collected from a Japanese paddy field during the flooded rice cultivation period. The g23 genes in these samples were quite distinctive in sequence from those obtained from marine environments. Phylogenetic analysis showed that most of g23 sequences belonged to two novel subgroups of T4-type phages (Paddy Soil subgroup and Rice Straw subgroup), although some of them were distantly related to well-studied subgroups of T4-type phages, for example, exoT-evens, T-evens and Groups II, III and IV of marine clones. This finding strongly indicates that the virus communities in soil are different from those in marine environments and that soil environments store novel structural and functional genes of viral origins.

References

- Common-types-fungi-found-soil: sciencing.com, Retrieved 31 July, 2019
- Mycorrhizae, NaturalHistoryOfFungi: nbm-mnb.ca, Retrieved 7 May, 2019
- The-role-of-soil-fungus: compostjunkie.com, Retrieved 4 February, 2019
- Viruses-soils, virus-research, cell-and-molecular-sciences: hutton.ac.uk, Retrieved 1 May, 2019

Soil Nematodes and Protozoa

Nematodes are non-segmented worms. They are beneficial to the soil as they play a vital role in controlling disease and cycle nutrients. Protozoa are eukaryotic organisms that primarily feed on bacteria and release excess nitrogen in the process. The chapter closely examines the key functions of nematodes and protozoa in the soil as well as their use for agricultural purposes.

Soil Nematodes

Nematodes are non-segmented worms typically 1/500 of an inch (50 μm) in diameter and 1/20 of an inch (1 mm) in length. Those few species responsible for plant diseases have received a lot of attention, but far less is known about the majority of the nematode community that plays beneficial roles in soil.

An incredible variety of nematodes function at several trophic levels of the soil food web. Some feed on the plants and algae (first trophic level); others are grazers that feed on bacteria and fungi (second trophic level); and some feed on other nematodes (higher trophic levels).

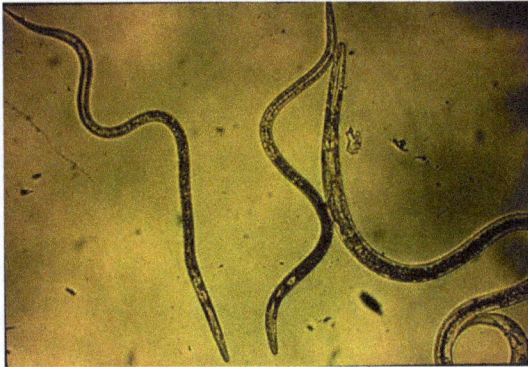

Most nematodes in the soil are not plant parasites. Beneficial nematodes help control disease and cycle nutrients

Free-living nematodes can be divided into four broad groups based on their diet:

- Bacterial-feeders consume bacteria.
- Fungal-feeders feed by puncturing the cell wall of fungi and sucking out the internal contents.

- Predatory nematodes eat all types of nematodes and protozoa. They eat smaller organisms whole, or attach themselves to the cuticle of larger nematodes, scraping away until the prey's internal body parts can be extracted.

- Omnivores eat a variety of organisms or may have a different diet at each life stage. Root-feeders are plant parasites, and thus are not free-living in the soil.

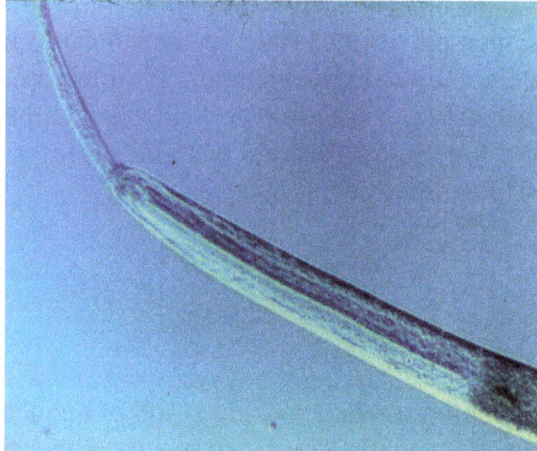

A predatory nematode consumes a smaller nematode.

Role of Nematodes

Nutrient cycling: Like protozoa, nematodes are important in mineralizing, or releasing, nutrients in plant-available forms. When nematodes eat bacteria or fungi, ammonium (NH_4^+) is released because bacteria and fungi contain much more nitrogen than the nematodes require.

Grazing: At low nematode densities, feeding by nematodes stimulates the growth rate of prey populations. That is, bacterial-feeders stimulate bacterial growth, plant-feeders stimulate plant growth, and so on. At higher densities, nematodes will reduce the population of their prey. This may decrease plant productivity, may negatively impact mycorrhizal fungi, and can reduce decomposition and immobilization rates by bacteria and fungi. Predatory nematodes may regulate populations of bacterial-and fungal-feeding nematodes, thus preventing over-grazing by those groups. Nematode grazing may control the balance between bacteria and fungi, and the species composition of the microbial community.

Dispersal of microbes: Nematodes help distribute bacteria and fungi through the soil and along roots by carrying live and dormant microbes on their surfaces and in their digestive systems.

Food source: Nematodes are food for higher level predators, including predatory nematodes, soil microarthropods, and soil insects. They are also parasitized by bacteria and fungi.

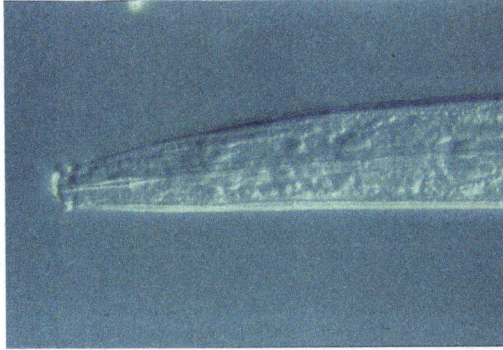

Fungal-feeding nematodes have small, narrow stylets, or spears, in their stoma (mouth) which
they use to puncture thecell walls of fungal hyphae and with draw the cell fluid.
This interaction releases plant-available nitrogen from fungal biomass

Disease suppression and development. Some nematodes cause disease. Others consume disease-causing organisms, such as root-feeding nematodes, or prevent their access to roots. These may be potential biocontrol agents.

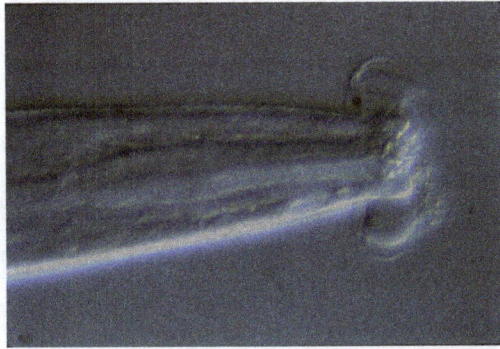

This bacterial-feeding nematode, Elaphonema, has ornate lip structures that distinguish it from other
nematodes. Bacterial-feeders release plant-available nitrogen when they consume bacteria.

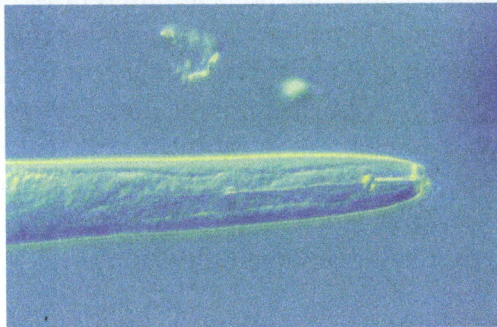

The Pratylenchus, or lesion nematode, has a shorter, thicker stylet in
its mouth than the root feeder below.

Nematodes are concentrated near their prey groups. Bacterial-feeders abound near
roots where bacteria congregate; fungal-feeders are near fungal biomass; root-feeders
are concentrated around roots of stressed or susceptible plants. Predatory nematodes
are more likely to be abundant in soils with high numbers of nematodes.

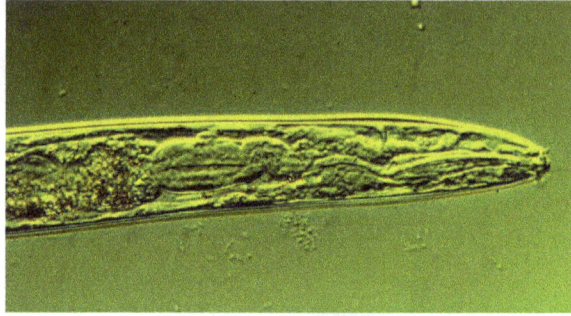
Root-feeding nematodes

Root-feeding nematodes use their stylets to puncture the thick cell wall of plant root cells and siphon off the internal contents of the plant cell. This usually causes economically significant damage to crops. The curved stylet seen inside this nematode is characteristic of the genus Trichodorus.

Because of their size, nematodes tend to be more common in coarser-textured soils. Nematodes move in water films in large (>1/500 inch or 50 μm) pore spaces.

Agricultural soils generally support less than 100 nematodes in each teaspoon (dry gram) of soil. Grasslands may contain 50 to 500 nematodes, and forest soils generally hold several hundred per teaspoon. The proportion of bacterial-feeding and fungal-feeding nematodes is related to the amount of bacteria and fungi in the soil. Commonly, less disturbed soils contain more predatory nematodes, suggesting that predatory nematodes are highly sensitive to a wide range of disturbances.

Nematodes and Water Quality

Nematodes may be useful indicators of soil quality because of their tremendous diversity and their participation in many functions at different levels of the soil food web. Several researchers have proposed approaches to assessing the status of soil quality by counting the number of nematodes in different families or trophic groups. In addition to their diversity, nematodes may be useful indicators because their populations are relatively stable in response to changes in moisture and temperature (in contrast to bacteria), yet nematode populations respond to land management changes in predictable ways. Because they are quite small and live in water films, changes in nematode populations reflect changes in soil microenvironments.

Nematode Trappers

One group of fungi may be a useful biological control agent against parasitic nematodes. These predatory fungi grow through the soil, setting out traps when they detect signs of their prey. Some species use sticky traps, others make circular rings of hyphae to constrict their prey. When the trap is set, the fungi put out a lure, attracting nematodes that are looking for lunch. The nematode, however, becomes lunch for the fungus.

Benefits of Nematodes in Healthy Soil Ecosystems

Although the threat of plant-parasitic nematodes damaging your crops is a concern, if a few important agricultural principles are followed – such as carefully designed crop rotations and building soil organic matter content – nematode-induced yield losses can be drastically reduced, while supporting nematode populations that will actually benefit your crops.

Nematodes are most abundant in the upper-most soil horizons, where up to 10 million individual nematodes can live per 10 square feet of soil. They subsist mostly in water-filled pore space near organic matter and plant roots.

Because of the havoc they can cause, plant parasitic nematodes, such as root-knot nematodes, have been the species most widely studied by scientists. But there are many other types of nematodes less studied, generally classified by their mouthparts and diet. Some strictly feed on either fungi or bacteria, while others are predatory, relying on other nematodes or protozoa for their diet; and others are omnivorous, able to feed on fungi and bacteria when their preferred prey is scarce, or conditions are unfavorable.

Roots of a pepper plant infected by southern root-knot nematodes
(Meloidogyne incognita) have extensive gall damage

Different types of nematodes play different roles in a soil system. In row crop systems, maintaining a diversified food web through soil conservation and organic matter additions can support nematode populations that actually enhance nutrient mineralization and plant nutrient availability. This is especially beneficial in farming systems reliant on organic nutrient sources, as bacterial feeding nematodes consume nitrogen-containing bacteria and release excess nitrogen as plant available ammonium (NH_4^+). Nematodes can also rejuvenate old bacterial and fungal colonies and spread these microorganisms into organic residues whose nutrients may otherwise remain immobile and unavailable to plants.

Light tillage, especially coupled with soil conservation practices and use of cover crops, can increase organic residue decomposition by bacteria and bacterial feeding nematodes, leading to more plant available nutrients for the season.

Although overly intensive tillage can disturb the soil food web, properly managed tillage can actually promote healthy soil ecosystems. Light soil disturbances – especially coupled with compost and manure additions – increase the availability of organic residues to be consumed by bacteria, which in turn stimulate bacterial feeding nematodes, leading to a net increase of available nitrogen for plant uptake. And although fungal feeding nematodes are more abundant in no-till and perennial agricultural systems, bacterial feeding nematodes are better at releasing plant available nitrogen than their fungal feeding counterparts.

Another type of nematode that can be beneficial to farming systems is the insect-parasitic bacterial feeding nematode. These nematodes, such as the species Heterorhabditis bacteriophora and Steinernema scapterisci, have mutualistic relationships with bacteria, in that both the nematode and bacteria rely on one another to reproduce and grow. In their infective juvenile stage, these specialized nematodes carry within their intestines specific bacteria. The nematodes can penetrate the body of many insect hosts, such as cutworms, mole crickets, citrus weevils, sawfly and fungus gnat larvae, and many more depending on nematode species, and release the bacteria into the insect's body cavity where it multiplies to the point of killing the insect. This allows the nematode to develop into an adult inside the insect's body and reproduce new juveniles, which emerge from the cadaver to search for a new host. Thousands of nematodes can be produced from just one infected insect host. These types of nematodes are even available commercially as a biological insect control, most commonly applied to moistened fields as liquid suspensions at a rate of about one million per acre, depending on the crop. As they are living organisms, care must be taken not to kill the nematodes with excessive pressure, temperature, agitation, or sun exposure. It is also important to select the correct nematode species to match target insect pests.

Thousands of Heterorhabditis bacteriophora insect-parasitic nematodes emerging out
of a wax moth cadaver, ready for use as a biological control to protect crops from
pests such as weevils, beetles, and flies

Predatory nematodes have biological control capabilities as well, in that they can regu-late populations of other nematodes – bacterial and fungal feeding – and most impor-tantly, root-eating plant parasitic nematodes. And as part of any healthy soil food web, there are a wide variety of natural enemies that help keep in check nematode popula-tions, such as predatory micro-arthropods and nematode ensnaring fungi. Agricultural systems designed to support a healthy soil ecosystem can therefore more successfully defend against plant parasitic nematodes and other crop diseases and pests. They can also facilitate enhanced nutrient cycling, which supports plant nutrient uptake, leading to overall healthy crop growth.

A white grub larva infected by a Heterorhabditis bacteriophora insect-parasitic
nematode next to two healthy white grub larvae for comparison.

Soil Nematodes in Organic Farming Systems

Most research on soil nematodes has focused on the plant-parasitic nematodes that attack the roots of cultivated crops. Less attention has been given to nematodes that are not plant-feeders and play beneficial roles in the soil environment.

Nematode Feeding Habits

Nematodes can be classified into functional groups based on their feeding habits, which can often be deduced from the structure of their mouthparts. In agricultural soils, the most common groups of nematodes are the bacterial-feeders, fungal-feeders, plant parasites, predators, and omnivores. Predatory nematodes feed on protozoa and other soil nematodes. Omnivores feed on different foods depending on environmental conditions and food availability; for example, omnivorous nematodes can be predators, but in the absence of their primary food source, they can feed on fungi or bacteria.

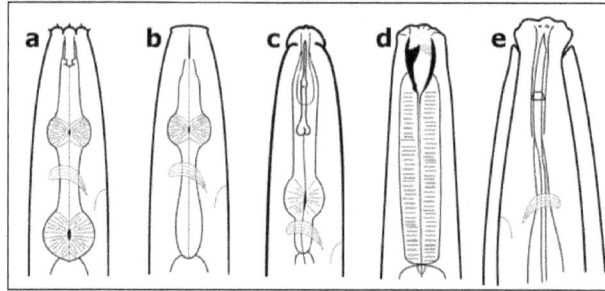

Nematodes can be classified into different feeding groups based on the structure of their mouthparts. (a) bacterial feeder, (b) fungal feeder, (c) plant feeder, (d) predator, (e) omnivore.

Importance of Nematodes in Agricultural Systems

Nematodes contribute to a variety of functions within the soil system. In agricultural systems, nematodes can enhance nutrient mineralization and act as biological control agents.

Nematodes and Soil Fertility

Soil nematodes, especially bacterial- and fungal-feeding nematodes, can contribute to maintaining adequate levels of plant-available N in farming systems relying on organic sources of fertility. The process of converting nutrients from organic to inorganic form is termed mineralization; mineralization is a critical soil process because plants take up nutrients from the soil primarily in inorganic forms. Nematodes contribute directly to nutrient mineralization through their feeding interactions. For example, bacterial-feeding nematodes consume N in the form of proteins and other N-containing compounds in bacterial tissues and release excess N in the form of ammonium, which is readily available for plant use. Indirectly, nematodes enhance decomposition and nutrient cycling by grazing and rejuvenating old, inactive bacterial and fungal colonies, and by spreading bacteria and fungi to newly available organic residues. In the absence of grazers, such as nematodes and protozoa, nutrients can remain immobilized and unavailable for plant uptake in bacterial and fungal biomass.

Bacterial-feeding nematodes are the most abundant nematode group in agricultural soils. Their abundance closely follows that of bacterial populations, which tend to increase when soil disturbances, such as tillage, increase the availability of readily-decomposable organic matter. Nitrogen mineralization in the soil occurs at a higher rate when bacterial-feeding nematodes are present than when they are absent. The contribution of bacterial-feeding nematodes to soil N supply depends, in part, on the quality and quantity of soil organic matter fueling the system. Net N mineralization from decomposing organic residues takes place when the carbon:nitrogen (C:N) ratio of organic residue is below 20 (that is, 20 parts C to 1 part N). When the C:N ratio is greater than 30, the rate of mineralization decreases because microbes compete for N to meet their nutritional requirements. In this situation, N is immobilized in the microbial biomass. Incorporation of manure, compost, and cover crops with intermediate C:N ratios (ranging from 10 to 18) may stimulate bacterial growth and the abundance of bacterial-feeding nematodes, and increase soil N availability to plants.

Fungal-feeding nematodes are relatively more abundant in less-disturbed (e.g. notill systems) and perennial systems, where conditions for fungal growth are promoted, than in disturbed systems. Like bacterial feeding nematodes, fungal-feeding nematodes contribute to the process of nutrient mineralization by releasing N and other plant nutrients from consumed fungal tissue. However, in agricultural systems, bacterial-feeding nematodes typically release more inorganic N than fungal-feeding nematodes.

Nematodes as Natural Enemies and Biological Control Agents

Predatory nematodes are of interest because of their role in regulating the populations of other organisms. They generally feed on smaller organisms like protozoa and other nematodes. Thus they can help moderate population growth of bacterial- and fungal-feeding nematodes and protozoa, and help regulate populations of plant-parasitic nematodes.

Insect-parasitic nematodes are species of bacterial-feeding nematodes that live in close association with specific species of bacteria; together, they can infect and kill a range of insect hosts. The infective juvenile stage of insect-parasitic nematodes seeks out insect hosts to continue its development into adults. Once a host is found, the nematodes penetrate the insect body and release their bacterial associates into the insect's body cavity. These bacteria multiply and overwhelm the immune response of the host insect, ultimately killing the host. The nematodes feed on these bacteria, mature, and reproduce until all the resources within the insect host are consumed; then, infective juvenile nematodes escape the insect host's body and disperse in the soil to seek new hosts. Insect-parasitic nematodes are available commercially for use in inundative releases to manage the populations of a variety of insect pests.

Plant-Parasitic Nematodes

Most plant-parasitic nematodes feed on the roots of plants. Some species attach to the outside surface of plant roots, piercing the root tissue to suck up the cellular content; other species pierce and penetrate the roots of plants, living and reproducing entirely within the root itself. A relatively small number of important plant-parasitic nematode species are known to cause substantial economic damage in cropping systems around the world. The determination of tolerance limits or economic thresholds for plant-parasitic nematodes varies with many factors like species, plant tolerance, and soil type. Because plant parasitic nematodes show varying degrees of host specificity, carefully designed crop rotations are usually a powerful tool for reducing nematode-associated yield losses.

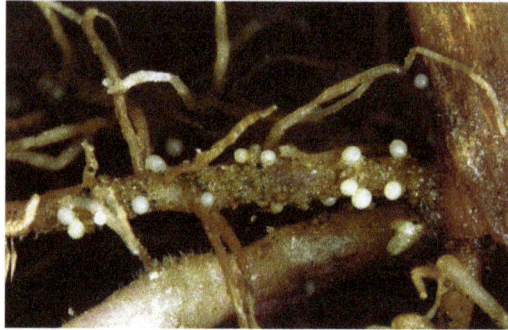

White potato cyst nematode, Globodera pallida (Stone) Behrens, on plant roots. Cyst nematode females attach to root systems with their mouthparts to feed, and then their bodies swell into egg-filled cysts that can be visible to the naked eye.

Soil Nematode Communities

The proportions of the different feeding groups in the soil nematode community vary between systems and seasons, and they are influenced by a variety of factors, including crop and soil management practices and the presence and abundance of natural enemies. Management practices like tillage, crop rotation, and the use of organic amendments influence the physical and biological characteristics of the soil that influence the abundance of nematodes. Fungal-feeding, predatory, and omnivorous nematodes are very sensitive to soil disturbances, and agricultural systems with fewer physical and chemical disturbances, such as pastures, hay fields, and orchards, tend to support larger populations of these nematodes than more frequently disturbed systems like vegetable- and row-crop fields. On the other hand, tillage and incorporation of organic residues increase the proportion of some bacterial-feeding nematodes, often offsetting declines in the numbers of other feeding groups and increasing the total abundance of nematodes. The wide variety of natural enemies that feed on or infect nematodes—predatory nematodes, predatory microarthropods, and nematode-trapping fungi, for example—may have a considerable impact on nematodes in agricultural systems.

Implications for Farming System Management

Agricultural management may increase the abundance of soil nematodes, primarily through the increase in the abundance of bacterial-feeding nematodes associated with tillage and the incorporation of organic residues. Soil conditions in agricultural production systems can be improved by enhancing nutrient availability and providing habitat for beneficial soil organisms. Maintenance of large populations of bacterial-feeding nematodes with practices that promote N mineralization throughout the growing season may enhance crop productivity, but a surplus of mineral N is not desirable from the environmental point of view because of an increased risk of nitrate leaching. In an ideal production system, N supply would be synchronized with plant demand. On the other hand, cultural practices like tillage or cultivation may reduce the complexity of the soil food web. Thus, a decrease in the frequency and intensity of tillage may promote the conservation of predatory nematodes and contribute to improved farming system performance.

Chemical Control of Nematodes

Plant-parasitic nematodes are at their most vulnerable during their active phase in soil when searching for the roots of host plants. Once endoparasitic species have penetrated a root, control with chemicals is more difficult as nematicidal compounds have to be non-phytotoxic and preferably systemic.

A nematicide that can be safely applied to growing plants and is translocated to the roots in sufficiently large amounts to kill endoparasitic or ectoparasitic nematodes has not been discovered. Oxamyl, a systemic compound that is translocated basipetally, is the only commercial product that is used as a foliar treatment, but its use as a liquid formulation is restricted in many countries for toxicological reasons.

There are several nematicides that can be used effectively for nematode pests of annual crops, but there appears to be little prospect for management of nematodes in many susceptible perennial crops without repeated application of nematicides. Only in certain cases will such treatments be justified economically. Since the discovery and wide-scale use of fumigant nematicides 50 to 60 years ago, a number of products and formulations have been developed for use against several nematode pests, and these are available in most regions of the world. Only in comparatively recent times have the dangers associated with the manufacture and use of these products become apparent. This has resulted in restrictions on use and sometimes withdrawal from the market. It seems that the age of the traditional fumigants and nematicides has passed, and the opportunity for managing nematodes with synthetic chemicals with broad biocidal activity is declining.

The development of new classes of nematicides with novel activity, that are effective when used in soil or applied directly to crops and are environmentally benign and specific to target pests, is perhaps an idealistic hope. Such compounds will, by definition, require considerable research effort, and if they are specific only to certain nematodes are unlikely to be considered an economic proposition for the traditional agrochemical company.

Chemical Control of Nematodes

The Demand

The elimination of nematodes from some crops is essential for certain export requirements, particularly of high-value horticultural products. Chemical treatment with fumigants or nematicides may be the only technique available, and from the plant quarantine standpoint it is important that their use is retained.

The use of chemicals in protected cropping may still be preferable to other techniques such as steam treatment for economic and practical reasons. The use of soil-less growing media in some north European countries has resulted in a decreased demand for chemical treatments. In southern Europe, the Mediterranean region and North Africa, many horticultural and salad crops are grown in soil under polythene and soil treatment with methyl bromide, dazomet or non-fumigant nematicides is widely practised. The cost of such treatments may be as much as US $500 per hectare per year and can only be justified if the crops are of high market value.

Efficacy

Fumigants: Several general purpose fumigants give excellent control of nematodes in soil. The efficacy is related to their high volatility at ambient temperatures. All fumigants have low molecular weights and occur as gases or liquids. As they volatilize, the gas diffuses through the spaces between soil particles; nematodes living in these spaces are killed.

Table: Nematicides currently available on world markets:

Chemical name	Trade name	Formulation
Fumigants		
Methyl bromide	Dowfume	Gas
1,3 dichloropropene	Telone/DD-95	Liquid
Ethylene dibromide[1]	Dowfume W-85	Liquid
Metam-sodium	Vapam	Liquid
Dazomet	Basamid	Dust (prill)
Methyl isothiocyanate	Di-Trapex	Liquid
Chloropicrin[1]	Larvacide	Liquid
Organophosphates		
Thionazin	Nemafos	Granular or emulsifiable liquid
Ethoprophos	Mocap	Granular or emulsifiable liquid
Fenamiphos	Nemacur	Granular or emulsifiable liquid
Fensulfothion	Dasanit	Granular
Terbufos	Counter	Granular
Isazofos	Miral	Granular or emulsifiable liquid
Ebufos	Rugby	Granular or emulsifiable liquid

Carbamates		
Aldicarb	Temik	Granular
Aldoxycarb	Standak	Flowable
OxamyI	Vydate	Granular or emulsifiable liquid
Carbofuran	Furadan/Curaterr	Granular or flowable
Cleothocarb	Lance	Granular

Use restricted:

Fumigants perform best in soils that do not have high levels of organic matter (which deactivates the toxicant) and that are free-draining but have adequate moisture. In general, fumigants are most effective in warm soils (12° to 15 °C) as dispersion is temperature related.

Methyl bromide, the most dangerous of the fumigants still in common use, has to be applied beneath a polyethylene sheet. In some countries this is done with specialized machinery that treats and covers the soil in one operation. The cover is removed some days later and the crop is sown or planted when all traces of the fumigant have dispersed.

Examples of recommended nematicidal dosages and treatments for some important crops:

Crop	Nematode pest	Nematicide	Application rate[2]	Application techniques
Potato	Globodera spp.	Aldicarb	2.24-3.36	Incorporated in row
		Oxamyl	4.0-5.5	
		Carbofuran	4.0-5.5	
Tomato, cucurbits	Meloidogyne spp.	Aldicarb	3.36	Incorporated in 30-cm bands
		Ethoprophos	0.9-2.9	Incorporated in bands
		OxamyI	0.6-1.2	Incorporated in bands
		Fenamiphos	1.6-3.3	Incorporated in bands
		Dazomet	30-50 g/m^2	Incorporated in bands and irrigated Time interval before planting
Citrus	Tylenchulus semipenetrans	Fenamiphos	10.8-21.6	Annual treatment applied along drip-line
		Aldicarb	5.5-11.0	Annual treatment applied along drip-line
Grape	Meloidogyne spp.	Fenamiphos (e.c. formulation)	10.0	In bands for nursery use
	Xiphinema index	Aldicarb	5-10	In bands for nursery use

Banana	Radopholus similis and/or	Carbofuran	2-4 ga.i. per plant	Applied around plant 2-3 times per year
	Helicotylenchus multicinctus and/or	Ethoprophos	2-4 ga.i. per plant	Applied around plant 2-3 times per year
	Pratylenchus spp. and/or	Fenamiphos	2-4 ga.i. per plant	Applied around plant 2-3 times per year
	Meloidogyne spp.	Isazofos	2-4 ga.i. per plant	Applied around plant 2-3 times per year
		Ebufos	2-4 ga.i. per plant	Applied around plant 2-3 times per year

1. Information taken from literature. Products may be unavailable for use in some countries. Economic and environmental justification should be evaluated before use. The omission of compounds does not imply that they are not equally suitable for nematode control.

2. kg a.i./ha unless otherwise stated.

Liquid fumigants EDB, metam-sodium and 1,3-D are applied to soil that has been prepared for planting. The soil surface is compacted with a roller after treatment which helps to seal the fumigant in the soil. Compounds releasing methyl isothiocyanate (dazomet, metam-sodium) work best in soils at >15 °C. In cooler soils, the period between treatment and planting may have to be extended to allow sufficient time for the product to disperse.

The liquid fumigant DBCP is the only volatile compound that can be applied to growing plants without causing phytotoxicity. However, its manufacture has now ceased for toxicological reasons and its use is banned in many countries.

Non-volatile nematicides. A number of organophosphate and oximecarbamate nematicides were developed in the 1960s, which had the advantage that application was relatively simple. As a consequence nematicide use became more widely practised. These compounds are active at dosages of 2 to 10 kg a.i./ha which are smaller than the 200 to 300 litres/ha required for treatment with liquid fumigants. Most of the early formulations of these products were as granules that, when applied to the soil surface (or preferably incorporated in the top 10 cm of soil), release the active ingredient, which is spread through the soil by rainfall or irrigation. The efficacy of soil penetration depends on the amount of moisture, organic matter and soil structure. Heavy soils with relatively small pore spaces are more difficult to treat than sandy soils which have larger pore sizes. Some chemicals, particularly the organophosphates, are absorbed in organic matter, in which case efficacy may be impaired.

In general, distribution of the active ingredient or its toxic degradation products is less efficient than that of fumigants and results with granular nematicides have sometimes been inconsistent. To be effective, nematicides have to persist long enough for nematodes to be exposed to lethal concentrations, which may be as low as 1 to 2m g/litre. Extended persistence is, however, not desirable if there is a risk of residues in the crop or the active compounds contaminating groundwater.

Persistence of soil-applied nematicides depends on the soil characteristics. In warm countries, relatively high soil temperatures may accelerate the natural degradation of nematicides, and in protected crops where even higher soil temperatures than out-of-doors may occur, the effective life of a nematicide might be as short as one to three weeks. The repeated use of products of similar structure can lead to the selection of a soil microflora that metabolizes these compounds and decreases their persistence.

Side-effects

All nematicides are eventually degraded if they remain in the topsoil where there is greatest microbial activity. Once nematicides or their degradation products are flushed through the upper soil layers their persistence may be extended. It is the problem of toxic products in groundwater that has led to the prohibition of fumigant and non-fumigant nematicides in some countries. The permitted level of pesticide residues in drinking-water in the European Union is 0.1m g/litre. In regions of intensive agricultural production these tolerance levels may be exceeded at certain times of the year.

Nematicides are highly toxic compounds that have very low LD50 values. This is particularly important for operators of application machinery and people at risk from exposure to the chemicals during their application. The liquid formulations of some of the non-fumigant nematicides are emulsifiable concentrates. Their use should therefore be restricted to skilled operators who take adequate safety precautions. This may not always be the case where basic levels of education are poor or where operators cannot read the instructions on the labels of the products. The application of nematicides to crops too near to harvest is another risk which pesticide residue monitoring may not be sufficiently well coordinated to prevent.

The incidence of pesticide poisoning and mortality in some countries serves as a grim warning of the risks that arise when pesticides are widely used under poor management.

The development cost of new products is more than US$20 million and the costs for registering these products are increasing as the criteria for their use are tightened. Conventional compounds (organophosphate or oximecarbamates) are unlikely to be developed if their toxicities are high.

New classes of nematicidal compounds are constantly being sought but there are currently nopromising materials close to commercial development. Avermectins, which are of microbial origin, have been developed for veterinary use and are powerful anthelmintics. Their efficacy against plant-parasitic nematodes is well established, however, because the compounds are complex they cannot be used successfully as soil treatments.

Soil Protozoa

Protozoa are single-celled animals that feed primarily on bacteria, but also eat other protozoa, soluble organic matter, and sometimes fungi. They are several times larger than bacteria - ranging from 1/5000 to 1/50 of an inch (5 to 500 µm) in diameter. As they eat bacteria, protozoa release excess nitrogen that can then be used by plants and other members of the food web.

Protozoa are classified into three groups based on their shape: Ciliates are the largest and move by means of hair-like cilia. They eat the other two types of protozoa, as well as bacteria. Amoebae also can be quite large and move by means of a temporary foot or "pseudopod." Amoebae are further divided into testate amoebae (which make a shell-like covering) and naked amoebae (without a covering). Flagellates are the smallest of the protozoa and use a few whip-like flagella to move.

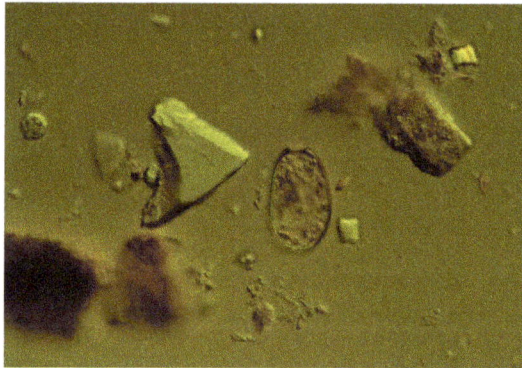

Protozoa play an important role in nutrient cycling by feeding intensively on bacteria. Notice the size of the speck-like bacteria next to the oval protozoa and large, angular sand particle.

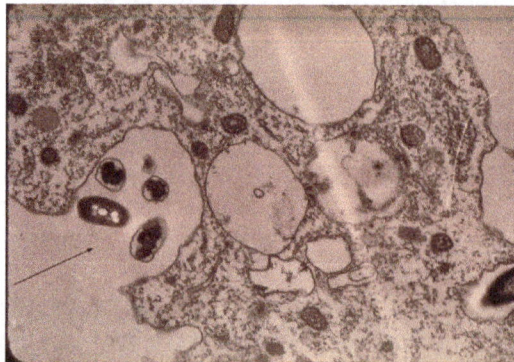

Bacteria ingested by an amoeba.

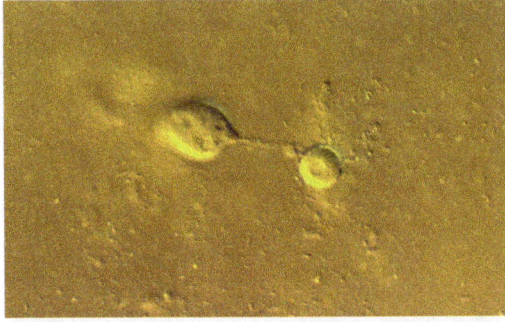

Flagellates have one or two flagella which they use to propel or pull their way through soil. A flagellum can be seen extending from the protozoan on the left. The tiny specks are bacteria.

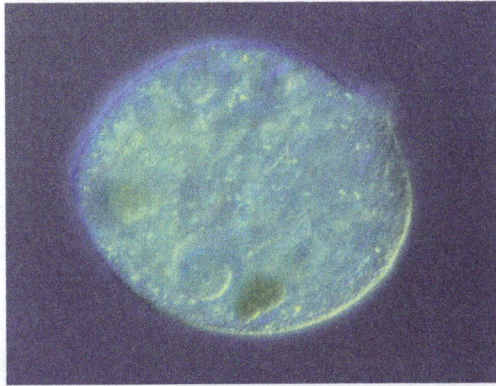

Ciliates are the largest of the protozoa and the least numerous. They consume up to ten thousand bacteria per day, and release plant available nitrogen. Ciliates use the fine cilia along their bodies like oars to move rapidly through soil.

Working of Protozoa

Protozoa play an important role in mineralizing nutrients, making them available for use by plants and other soil organisms. Protozoa (and nematodes) have a lower concentration of nitrogen in their cells than the bacteria they eat. (The ratio of carbon to nitrogen for protozoa is 10:1 or much more and 3:1 to 10:1 for bacteria.) Bacteria eaten by protozoa contain too much nitrogen for the amount of carbon protozoa need. They release the excess nitrogen in the form of ammonium (NH_4^+). This usually occurs near the root system of a plant. Bacteria and other organisms rapidly take up most of the ammonium, but some is used by the plant.

Another role that protozoa play is in regulating bacteria populations. When they graze on bacteria, protozoa stimulate growth of the bacterial population (and, in turn, decomposition rates and soil aggregation.) Exactly why this happens is under some debate, but grazing can be thought of like pruning a tree - a small amount enhances growth, too much reduces growth or will modify the mix of species in the bacterial community.

Protozoa are also an important food source for other soil organisms and help to suppress disease by competing with or feeding on pathogens.

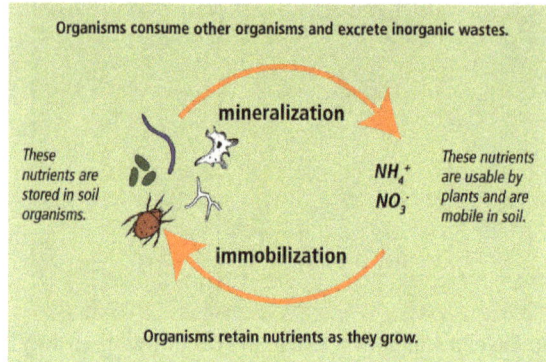

Organisms consume other organisms and excrete inorganic wastes.

mineralization

These nutrients are stored in soil organisms.

NH_4^+
NO_3^-

These nutrients are usable by plants and are mobile in soil.

immobilization

Organisms retain nutrients as they grow.

Mineralization and Immobilization

Protozoa need bacteria to eat and water in which to move, so moisture plays a big role in determining which types of protozoa will be present and active. Like bacteria, protozoa are particularly active in the rhizosphere next to roots.

Typical numbers of protozoa in soil vary widely - from a thousand per teaspoon in low fertility soils to a million per teaspoon in some highly fertile soils. Fungal-dominated soils (e.g. forests) tend to have more testate amoebae and ciliates than other types. In bacterial-dominated soils, flagellates and naked amoebae predominate. In general, high clay-content soils contain a higher number of smaller protozoa (flagellates and naked amoebae), while coarser textured soils contain more large flagellates, amoebae of both varieties, and ciliates.

Most protozoa eat bacteria, but one group of amoebae, the vampyrellids, eat fungi. The perfectly round holes drilled through the fungal cell wall, much like the purported puncture marks on the neck of a vampire's victim, are evidence of the presence of vampyrellid amoebae. The amoebae attach to the surface of fungal hyphae and generate enzymes that eat through the fungal cell wall. The amoeba then sucks dry or engulfs the cytoplasm inside the fungal cell before moving on to its next victim.

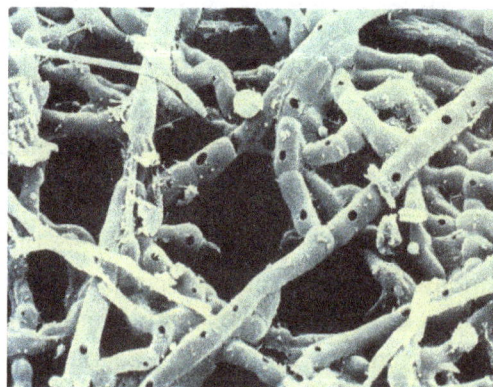

Fungi-eating protozoa

Vampyrellids attack many fungi including root pathogens, such as Gaeumannomyces graminis, shown in the photo. This fungus attacks wheat roots and causes take-all disease.

Soil Protozoa as Bioindicators

Heterotrophic protozoa have been ignored by general ecologists for a long time. Although the reason for this neglect is difficult to pinpoint, it was probably due to lack of expertise or because protozoa are more difficult to handle than larger metazoans. The situation changed rather abruptly about fifteen years ago, when marine biologists and later also freshwater limnologists recognized the importance of protozoa in the pelagic food chains and created the concept of the 'microbial loop', which suggests that a considerable fraction of the reduced carbon is mineralized before it reaches large plankton organisms such as copepods and fish. It is now clear that most of the bacterial production is consumed by small protozoa (mainly heterotrophic nanoflagellates, HNF), which in turn serve as food for larger protozoa (mainly ciliates) and small metazoa. This also applies for streambed sediments, in which 80 to 183% of the annual bacterial productivity is consumed by protozoa and protozoan production amounts to up to 30% of that of the macrozoobenthos.

The exciting results obtained by marine and freshwater ecologists also rejuvenated research on soil protozoa. Although the exact contribution of protozoa to the soil energy flux is still under discussion, it is clear that they enhance flows of nutrients in soil to the benefit of microorganisms, plants and animals. Estimates from several independent food web studies confirm that protozoa, on average, are responsible for about 70% of the soil animal respiration and for 14–66% and 20–40% of the C and N mineralization, respectively. Accordingly, decomposition is delayed by at least 33–39% (8–18 months) if soil animals are excluded from the decomposer cycle. Last but not least, protozoa play a significant role in earthworm nutrition and may stimulate bacterial antagonism, which in turn could provide the basis for a novel approach to the biological control of plant diseases.

Their dynamics and community structures should be powerful indicators of changes in the biotic and abiotic environment. In fact, protozoa have been widely used as indicators in various freshwater ecosystems since the turn of the century. In soil, protozoa are still rarely used as indicators, although reviews of the literature have catalogued about 200 pertinent papers demonstrating the usefulness of protozoa to indicate changes in the soil environment. Thus, the present review should stimulate applied soil ecologists to exploit more widely and successfully the indicative potential of soil protozoa.

Protozoa as Indicators of Soil Quality: Pros and Cons

Bioindicators are, in a broad ecological sense, organisms that can be used for detection and qualitative and/or quantitative characterization of a certain environmental factor or of a complex of environmental factors; a narrower definition confines bioindicators to human influences. Of course, indication is dependent on the overall biology of the organisms, especially on some basic physiological properties. As compared to the meso- and macrofauna, protozoa and small metazoans (i) consume more food per mass unit, (ii) have a higher respiration rate per mass unit, (iii) have shorter generation and (iv) life times, and (v) reproduce faster. Thus, the various groups of organisms have different indicative properties, i.e., are not, or usually are not, equivalent and replaceable. Accordingly, a well-designed bioindication study should contain at least one group each of the micro-, meso- and macrofauna. For instance, a comparative study on the effects of conventional and organic farming on soil animals revealed that testate amoebae were significantly reduced in the conventional plots, while earthworms remained unchanged.

Furthermore, indication must be 'directed', i.e., the indicator group(s) must be chosen according to the peculiarities of the habitat and the goal of the study, and not, as frequently recognizable, according to the facilities and expertise available. If, for instance, the effects of a certain fertilizer on the soil fauna of a spruce forest are to be investigated, then it will be crucial that protozoa, especially testate amoebae, are included in the set of bioindicators because they are highly abundant in such habitats, from which other potential bioindicators such as most earthworms are excluded due to the high acidity.

The pros and cons of using heterotrophic soil protozoa as bioindicators can be summarized as follows:

1. Protozoa are an essential component of soil ecosystems, because of their large standing crop and production. Changes in their dynamics and community structures very probably influence the rate and kind of soil formation and soil fertility.

2. Protozoa, with their rapid growth and delicate external membranes, can react more quickly to environmental changes than any other eukaryotic organism and can thus serve as an early warning system and excellent tool in bioassays. Results are obtained within 24 h, faster than with any other eukaryotic test system.

3. The eukaryotic genome of the protozoa is similar to that of the metazoa. Their reactions to environmental changes can thus be related to higher organisms more convincingly than those of the prokaryotes.

4. Morphologic, ecologic, and genetic differentiation among globally distributed protozoan species is low, at least in the few soil ciliates and testate amoebae studied so far suggesting that many indicator species can be used worldwide.

5. Protozoa inhabit and are particularly abundant in those soil ecosystems that almost or entirely lack higher organisms due to extreme environmental conditions, e.g. alpine regions above the timberline, Arctic and Antarctic biotopes.

6. Protozoa are not readily dislodged in soil. Many (but not all) are ubiquitous and are useful for comparing results from different regions. Differences in patterns of distribution are almost entirely restricted to passive vertical displacement;

thus, the difficult problem of horizontal migration, especially with the epigaeon, does not affect the investigations.

There are, however, several factors that have apparently restricted the use of soil protozoa and soil animals in general as bioindicators:

1. There is an immense number of species, many of which have not yet been described; more than 1000 may occur in a square metre of forest soil. Many specialists are needed for identification and each species has specific requirements that are often incompletely known.

2. Enumeration of soil organisms is difficult and time consuming.

3. Animals need other organisms for food. Thus, the constellation of factors is more complicated than in plants and indication often remains unspecific, i.e., different factors induce similar reactions.

4. Most soil organisms are inconspicuous and invisible to the naked eye, making them unattractive to many potential investigators.

5. There is a lack of comprehensive, easy-to-use identification literature.

Biology of soil Protozoa Related to their use as Bioindicators

Traditionally, soil protozoa are classified as naked amoebae, testate amoebae, flagellates, ciliates, or sporozoans. General morphology and ecology of the vari ous groups of protists are described in textbooks and the book edited by Darbyshire. Compilations on representative abundances in various soils and habitats are provided by Heal, Foissner and Cowling.

Naked Amoebae

About 60 species have been recorded to occur in soil. It is very likely that this is only a small portion of the species actually present. The 'naked amoebae' are not a monophyletic assemblage but belong to several phyla having different ecological properties. Many soil protozoologists consider the naked amoebae as the most important group of soil protozoa due to their often great abundance; numbers between 2000 and 2 000 000 g−1 dry mass of soil have been reported. However, all estimates are from dilution culture methods and thus highly questionable, especially regarding the active portion possibly, most are in an inactive (cystic) condition. Unfortunately, most soil naked amoebae are very small (<30 µm) and attached to soil particles. Thus, they cannot be counted in simple soil suspensions to obtain information about the active and cystic proportion. Certainly, some or many are active for at least some time. The common soil naked amoebae feed mainly and partially selectively on bacteria. With their tiny and flexible pseudopodia they can exploit micropores having a diameter of only 1 µm.

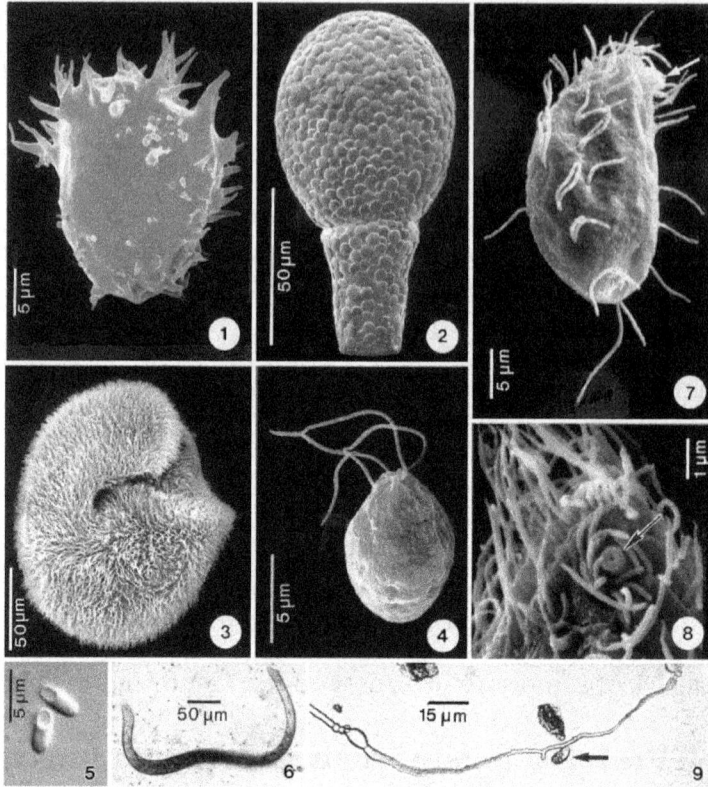

Main types of soil protozoa in the scanning electron microscope and light microscope. 1: Mayorella sp., a naked amoeba with cone-shaped pseudopodia mainly in the anterior body half; 2: Nebela vas, a Gondwanan testate amoeba. The shell is made of circular platelets originating from its prey, another testate amoeba, Trinema lineare; 3: Bresslauides discoideus, a large colpodid ciliate possibly restricted to terrestrial habitats of Laurasia; 4: Polytomella sp., a common heterotrophic soil flagellate; 5: Ciliatosporidium platyophryae, a microsporidian parasite affecting the soil ciliate Platyophrya terricola; 6: Circinella arenicola, a long and slender soil ciliate easily confused with small nematodes; 7–9: Pseudoplatyophrya nana (7, 9) and Grossglockneria acuta (8) are colpodid ciliates which evolved a minute, highly complicated feeding tube to penetrate the cell wall of yeasts and fungi. They cannot feed on other items.

Table: Comparison of population parameters for soil testaceans from different sites and biotopes.

Parameter	Moss under beech forest (moder)	Ash-maple forest (mull)	Alpine rendzina (moder)	Alpine mat (moder)	Aspen woodland (mor)	Beech forest (mull)
Annual mean density (×106 m-2)	1.7	33.6	39.6	20.1	261	84
Annual mean biomass (mg m-2)	15.5	1033	2209	1165	723	1715

Number of generations (per year)	16.0	12.5	n.d.	8	90	n.d
Mortality rate (% per day)	3.0	8.5	n.d.	n.d.	10.8	n.d.
Production numbers (×106 m-2)	145	940	29	118	90 930	358 000
Production biomass (g m-2)	0.11	25.9	1.4	5.4	206	73
Biomass turnover (PB/B)	8.1	43.9	0.7	5	285	43

However, there are also many mycophagous naked amoebae. Some of them feed by perforation lysis, producing discrete holes, 0.2 to 6 mm in diameter, in the fungal cell wall, like the mycophagous ciliates. There is evidence from pot experiments that such activities can reduce inoculum levels of plant pathogenic fungi in soils and can reduce the severity of take-all disease of wheat by Gaeumannomyces graminis var. tritici.

Testate Amoebae

This group comprises amoeboid organisms with a shell (test) either produced by the cell itself (often composed of silicium scales = idiosomes), or composed of foreign particles (e.g., sand grains = xenosomes) agglutinated to a membranous sheath. We now know that the testate amoebae are at least biphyletic: the idiosome-bearing, filose taxa (e.g., Euglypha) belong to the phylum Rhizopoda, whereas the xenosome-bearing, lobose forms (e.g., Difflugia) belong to the phylum Amoebozoa. This is in accordance with the often distinctly different autecology of filose and lobose testaceans. About 300 species and many varieties have been found in terrestrial environments, some of which have a restricted Laurasian or Gondwanan distribution, e.g., Nebela (Apodera) vas. Standing crop numbers are moderately high, with 100–1000 individuals per gram dry mass in mineral soils, 1000–10 000 in meadow topsoils and grasslands, and 10 000–100 000 in leaf litter. However, as many testate amoebae species are rather voluminous, their standing crop and production biomass often surpasses that of all other protozoans. Thus, their omission in some recent ecosystem studies is an unfortunate mistake.

In my opinion, testate amoebae are very useful indicator organisms in a wide range of terrestrial biotopes, not only because they are more easily counted and identified than the other soil protozoans, but also because of their high biomass and considerable abundance. Furthermore, they have a considerable species and lifestyle diversity and a distinct and deep vertical distribution. Accordingly, they can indicate a wide range of biotic and abiotic variables, so much the more in that rather detailed autecological data are available from most of the common species.

Flagellates

About 260 species have been recorded from soils worldwide, most were first or subsequently found in limnetic or coprozooic biotopes. However, new species are described rather steadily from soil, indicating that we possibly know only a small proportion of the species actually present. Like amoebae, flagellates are polyphyletic, showing a wide range of morphologies and ecologies. Many soil protozoologists consider the flagellates, together with the naked amoebae, as the most important group of soil protozoa due to their often great abundances and short generation times (usually <5 h under laboratory conditions); numbers between 0 and 10^6 g^{-1} of soil have been reported. However, as with the naked amoebae, most counts rely on culture techniques and MPN esti-mates which do not give an accurate indication of the active cells present. The few direct counts available indicate that most are in an inactive encysted condition. Ecologically, flagellates have much in common with the naked amoebae: most feed on bacteria, are small (< 20 μm), and have an amoeboid flexibility that allows them to inhabit even very small soil pores which cannot be exploited by larger protozoans, such as most testate amoebae and ciliates. For example, Hattori isolated flagellates from 80% of the 1–2 mm sized soil aggregates examined.

Table: Diversity of soil ciliates in Europe, Africa, Australia, and Antarctica

Characteristics	Europe (99 samples)	Africa (92 samples)	Australia (157 samples)	Antarctica (90 samples)
Total number of species	345	507	361	95
Species/sample (mean)	26	35	23	4
Undescribed species	185	240	154	14
Undescribed species (%)	54	47	43	15
Undescribed species/sample (mean)	1.9	2.6	1.0	0.2
Undescribed species/sample (%)	7.2	7.5	4.4	3.9

Ciliates

The ciliated protozoa comprise unicellular, heterokaryotic organisms having a macro-nucleus and a micronucleus of distinctly different size and function within the same cytoplasm. The macronucleus, which is usually highly polyploid, divides amitot-ically during asexual reproduction and controls mainly somatic functions (e.g. RNA synthesis). The diploid micronucleus is active mainly during sexual reproduction (conjugation). Diversity of ciliates is high in terrestrial habitats, that is, at least 2000 species, 70% of which have not yet been described. Most soil ciliates feed on bacteria (39%) or are predaceous (34%) or omnivorous (20%). Some, however, are strictly my-cophagous and highly characteristic for terrestrial habitats a few (mainly metopids) are anaerobic, providing a simple tool to assess the soil oxygen regime. Soil ciliates have a particular vertical distribution, which must be taken into account when they are used as bioindicators. In contrast to the testate amoebae, appreciable numbers of active cil-iates occur only in the uppermost litter layer, where abundances of up to 10 000 indi-viduals per gram dry mass of litter are reached. In the humus horizon and in mineral soils active ciliates are rare, although many cysts are present. Thus, habitats such as meadow topsoils and arable lands contain very few active ciliates, usually less than 100 individuals per gram dry mass of soil. The reasons for these distinct distributions are not yet known; obviously, excystment is suppressed under field conditions in evolved soils. Thus, the use of ciliates as bioindicators is usually confined to habitats with a distinct litter layer, e.g. forests or heavily disturbed soils, where ciliates are an excellent tool to monitor recovery of soil health following major soil pollution or disturbance.

Sporozoans

This group, which contains only parasitic organisms, has rarely been used as bioindicators. There are, however, at least three studies available, which suggest that they are very useful for this purpose. Pizl found a significantly increased infection of earthworms by monocystid gregarines when the earthworms were exposed to an herbicide, zeazin 50, for 26 weeks. The same was observed in roadside soils polluted by heavy metals. Purrini reported a dramatically increased infection of soil invertebrates with parasitic protozoa (gregarines, coccidids, microsporids) in regions with high SO_2 deposits. Certainly, these are quite interesting data, not only from the bioindicative point of view, but also because they indicate that protozoa may play a significant role in regulating densities of soil invertebrates. In fact, sporozoans also parasitize soil ciliates, albeit rarely.

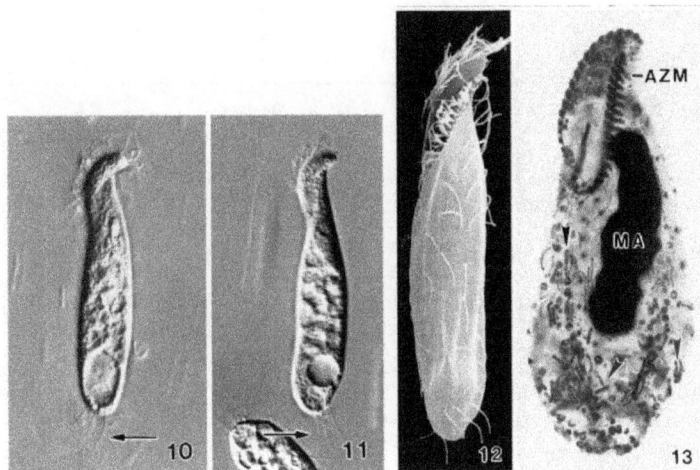

Metopus hasei, a common ciliate in insufficiently aerated soils in vivo (10, 11), in the scanning electron microscope (12), and after silver impregnation (13). 10–12: M. hasei is about 100mm long and has caudal cilia (arrows), which were not preserved in the SEM; 13: Metopid ciliates lack mitochondria and have many rod-shaped methanogenic bacteria in the cytoplasm (arrowheads). AZM: adoral zone of membranelles, MA: macronucleus.

Examples of using soil protozoa as bioindicators:

A few representative examples from these compilations and some very recent data and developments will be discussed in the following paragraphs. The examples emphasize the need for species identification, i.e., investigation not only of total individual and biomass changes, if results obtained with bioindicators are to be meaningful and re-producible.

Soil Protozoa as Indicators in Natural Ecosystems

Oxygen Regime

Certain ciliates are excellent and simple bioindicators for the soil's oxygen regime, es-pecially for periodic or sporadic oxygen depletion, which is often difficult to detect with conventional physicochemical methods. The indication is based on metopid hetero-trichs, which can live and reproduce only under microaerobic and/or anaerobic con-ditions because they lack typical mitochondria. Metopids live in symbiosis with meth-anogenic bacteria (usually Methanobacterium formicicium), using the hydrogen gas produced in the ciliates' hydrogenosomes, which are very likely highly modified mito-chondria. The most common soil metopid is Metopus hasei, a slender, medium-sized (70–120 µm) species. The methanogenic bacteria can be visualized by silver carbonate impregnation.

Table: Occurrence of metopid (anaerobic) ciliates in various soils and habitats.

Sites	Habitat	Soil depth investigated (cm)	No. of samples investigated	No. of metopid species
1	Beech forest in Salzburg, Austria	0–10	5	0
2	Beech forest in Lower Austria	0–10	10	0
3	Spruce forest in Upper Austria	0–3	48	0
4	Two bottomlands in Lower Austria	0–10	20	2
5	Two xerothermic, treeless sites within Site 4	0–10	20	0
6	Two ploughed fields near Sites 4 and 5	0–10	20	0
7	Alpine mat in Austria	0–10	8	0
8	Eutrophified alpine mat in Austria	0–10	3	1
9	Swamp soil near Cairns, Australia	0–10	1	6
10	Corn field near Site 9	0–10	1	0
11	Two floodplain rain forests in Manaus, Brazil	0–10	2	10
12	Two terra firma rain forest soils near Site 11	0–10	2	0
13	Municipal landfill sites in England	50–800	>10	1

Table shows a selection of habitats (soils) which are known to be usually aerobic (e.g., forest litter, ploughed fields) or anaerobic for some time (e.g., flood plains). Metopids occur only in the anaerobic soils. The eutrophic (by cattle dung and urine) pasture is of special interest, showing that indication by metopids is not trivial and could be helpful in detecting oxygen deficits that are not particularly obvious. Certainly, more detailed investigations are needed, but the available field evidence looks highly suggestive.

Differentiation of Humus Types

The important role of the humus type in the distribution of soil animals and for soil fertility is well known. Relevant data on soil protozoa were provided by several authors and reviewed by Foissner. Unfortunately, the subject has received little attention during the past decade.

The most valuable protozoan humus indicators are= several species of testate amoebae; for ciliates the data are still preliminary. Although the indicator species do not occur exclusively in a certain type of humus, their incidence and individual numbers are mostly low outside their preferred habitat. Even subtle differences in the humus type, which frequently pose problems to soil scientists, are nicely distinguished by the testate amoebae. Mull and mor are clearly distinguishable also by the relation of full (alive) and empty (dead) testaceans, whose shells decompose markedly slower in mor than in mull soils: ≤1 : 2 in mull, 1 : 2 to 1 : 10 in moder, and ≥1 : 10 in mor.

Soil Protozoa as Indicators in Ecosystems Under Human Influence

Pesticides

Petz and Foissner investigated the effects of a fungicide, mancozeb, and an insecticide, lindane, on the active microfauna of a spruce forest soil using a completely randomized block design and a direct counting method. The effects were evaluated 1, 7, 15, 40, 65 and 90 days after application of a standard or high (10x) dose. Mancozeb, even at the higher dose, had no pronounced acute or long-term effects on absolute numbers of the taxa investigated. The number of ciliate species decreased one day after treatment with the standard dose ($0.05 < P ≤ 0.1$) but soon recovered. However, the community structure of ciliates was still slightly altered after 90 days. Mycophagous ciliates (e.g. Pseudoplatyophrya nana) were substantially reduced in the first weeks after application of the fungicide. Testaceans were not reduced before Day 15 with the higher dose or before Day 40 with the standard dose ($0.05 < P ≤ 0.1$). The standard dose of lindane caused acute toxicity in ciliates and rotifers ($P ≤ 0.05$), although the latter soon recovered. The number and community structure of ciliate species were still distinctly altered after 90 days ($0.05 < P ≤ 0.1$), indicating the crucial influence of lindane. Testaceans were reduced only on Day 15 and nematodes only on Day 40 ($0.05 < P ≤ 0.1$). At the high dose of lindane, severe long-term effects occurred in soil moisture, total rotifers ($P ≤ 0.05$), total nematodes ($0.05 < P ≤ 0.1$) and in the structure of the ciliate community. Some species became more abundant 90 days after treatment with lindane, e.g., Colpoda inflata, C. steinii and Pseudoplatyophrya nana, possibly due to reduced competition and their r-selected survival strategy, whereas Avestina ludwigi, very dominant in the control plots, became extinct.

Table: Species characteristic of the testacean and ciliate communities in mull and mor soils.

Type of humus	Characteristic species	
	Testaceans	Ciliophora
Mull	Centropyxis plagiostoma	Urosomoida agilis
	Centropyxis constricta	Urosoma spp.
	Centropyxis elongata	Hemisincirra filiformis
	Plagiopyxis minuta	Engelmanniella mobilis
	Geopyxella sylvicola	Grossglockneria hyalina
	Paraquadrula spp.	Colpoda elliotti
Moder and mor	Trigonopyxis arcula	Frontonia depressa
	Plagiopyxis labiata	Bryometopus sphagni
	Assulina spp.	Dimacrocaryon amphileptoides
	Corythion spp.	Avestina ludwigi
	Nebela spp.	

Table: Dominance (%) of testacean species characteristic of mull and mor/moder in some alpine pseudogleys.

Sites	A	B	E	C	D
Type of humus	Mull-like moder			Moder	
C/N ratio	9.9	10.4	10.2	10.2	13.9
Plagiopyxis minuta	0.4	0.0	5.6	0.0	0.0
Centropyxis plagiostoma	0.4	0.0	2.3	0.0	0.0
Centropyxis elongata	1.8	6.0	0.9	0.4	0.0
Trigonopyxis arcula	0.0	0.0	0.0	1.1	1.6
Nebela parvula	0.0	0.0	0.0	0.0	1.9
Corythion spp.	0.0	0.0	0.5	1.4	1.3
Assulina spp.	1.3	1.7	0.0	2.2	4.5
Plagiopyxis declivis	22.2	14.8	17.9	12.7	7.2

Generally, there were marked differences between the effects of the standard and the high dose of lindane, but not with mancozeb. Ciliates showed very pronounced changes after the pesticide applications, whereas testaceans were more resistant.

Global Warming

Treonis and Lussenhop grew Brassica nigra at either ambient or twice-ambient CO_2 levels within open-top chambers in the field for 4 weeks. Plant biomass, above and below ground, was unaffected by increased CO_2. Direct count bacterial density was also unchanged under increased CO_2. Flagellate density tended to increase, whereas the number of naked amoebae significantly declined under increased CO_2. Treonis and Lussenhop suggested that these changes were caused by a trophic transfer of the increased CO_2 fertilization effect through the soil food chain. Unfortunately, the protozoan counts of Treonis and Lussenhop are based on a culture method which is known to give highly ambivalent results; thus, the study is not definitive. However, the data at least suggest that soil protozoa could be a valuable and efficient system for exploring the effects of increased CO_2 levels and hence global warming on soil organisms. Smith in fact predicted from a faunal censu and some experiments that the ciliate genus Colpoda, which has its southern limit of frequent and ubiquitous distribution at the sub-Antarctic/maritime Antarctic boundary, would quickly spread over the maritime Antarctic as a result of global warming.

Percentage of the dominant species of active ciliates and testate amoebae in spruce forest litter 1 and 90 days after treatment with mancozeb and lindane at normal (0.096 g m⁻² and 6 g m⁻² active ingredient, respectively) and high doses (0.96 g m⁻² and 60 g m⁻² active ingredient, respectively).

Species	Day	Control	Mancozeb		Lindane	
			1X	10X	1X	10X
Ciliates						
Avestina ludwigi Aescht and Foissner	1	23.5	44.3	37.4	15.2	0.0
	90	41.3	45.8	42.9	26.7	0.0
Platyophrya spumacola Kahl	1	19.4	12.7	21.7	12.1	0.0
	90	13.1	9.8	13.6	22.4	4.4
Pseudoplatyophrya nana (Kahl)	1	18.5	10.0	6.0	15.2	0.0
	90	10.8	13.6	17.7	24.2	35.0
Colpoda steinii Maupas	1	4.5	1.4	4.0	6.1	xd
	90	0.0	0.0	0.7	2.5	44.8
Colpoda inflata (Stokes)	1	0.1	1.0	0.5	0.0	0.0
	90	0.8	0.0	0.7	0.6	9.3
Testate amoebae						
Corythion dubium Taranek	1	48.1	74.6	63.3	46.0	51.3
	90	47.9	52.2	48.0	39.7	32.7
Trinema lineare Penard	1	11.3	7.5	0.0	18.7	12.2
	90	9.6	17.5	17.8	18.3	15.7
Schoenbornia humicola (Schönborn)	1	0.0	0.0	2.8	3.0	8.7
	90	9.6	9.7	6.4	6.9	17.0

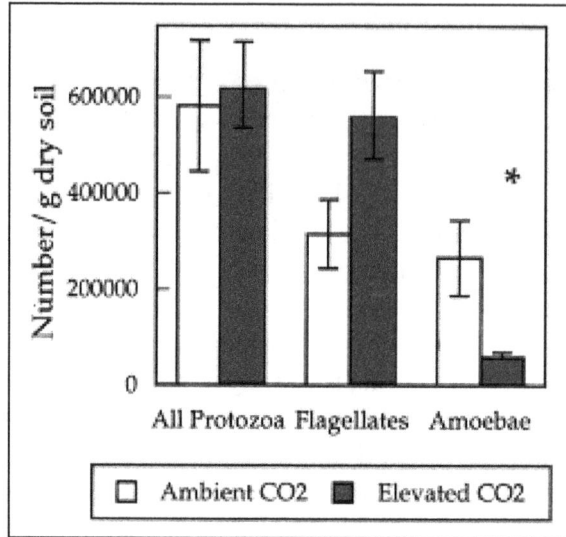

Response of 'total protozoa' (flagellates + naked amoebae), flagellates, and amoebae in soils of plants grown under elevated CO_2. Values were obtained with a culture method and show treatment means (n = 20); bar lengths represent 1 unit of standard error. Asterisk (*) indicates significant difference (P = 0.046) to control.

Forest Decline

The effects of slow release, organically enriched magnesite fertilizers on soil organisms of a declining spruce forest in Upper Austria were studied during a five-year period using a completely randomized block design. For revitalization, 2000 kg ha−1 BIO-MAG© (90% magnesite and 10% dried fungal biomass; 'biomag plots') and 3000 kg ha−1 BACTOSOL© (dried bacterial biomass) + 2000 kg ha−1 biomag ('bactosol-biomag plots') were applied once to an old and to a young stand. The treatments caused a mean pH rise of about 0.9 units. The biomasses of testaceans and the individual numbers of ciliates significantly decreased on some plots. Likewise, community structures were changed: individual numbers of acidophilic testacean species decreased and circumneutral species increased; mycophagous and bacteriovorous ciliates increased or decreased depending on species. Apart from the increase in pH, these changes could be related to altered numbers and kinds of food organisms such as fungi and bacteria, as indicated by an increased catalase and protease activity and decreased phosphatase and cellulolytic activity. The abundances of rotifers and proturans significantly increased in the bactosol-biomag plots of the old stand, and earthworm abundances increased with all treatments. Nematodes, mites, springtails, and enychtraeids were hardly affected. In general, soil fauna and enzyme activities were influenced to a greater extent by the bactosol-biomag than by the biomag treatment. These investigations also showed that it is essential to take into account species and their aute cology, i.e., indication remains rather unspecific and ambiguous when confined to abundance and biomass changes.

Abundance, individual dominance, and frequency of important ciliate groups and species in the litter layer (0–3 cm) of a declining, ameliorated spruce forest.

Feeding type	Parameter	Treatment						Difference to (%; C,c = 100%)			
		C	M	O	c	m	o	M	O	m	o
Fungal feeders (total)	ID%	55	58	62	62	53	58	5	13	−14	−6
Avestina ludwigi	Ind.	95	97	59	47	42	30	2	−38	−11	−36
	ID%	15	25	14	20	20	14	66	−5	2	−30
	F%	94	86	81	94	83	81	−9	−14	−12	−14
Rostrophryides australis	Ind.	88	38	62	38	24	24	−57	−30	−37	−37
	ID%	14	10	15	16	11	11	−30	7	−28	−31
	F%	100	92	96	88	83	90	−8	−4	−6	2
Grossglockneridaed	Ind.	55	34	47	17	24	38	−38	−15	41	124
	ID%	9	9	11	7	11	17	0	30	59	144
	F%	85	90	88	90	90	83	6	4	0	−8
Bacteria feeders (total)	ID%	24	22	17	14	25	25	−8	−29	79	79
Cyclidium muscicola	Ind.	126	39	43	27	23	16	−69	−66	−15	−41
	ID%	20	10	10	11	11	7	−50	−48	−4	−35
	F%	100	83	92	79	92	73	−17	−8	16	−8
Colpoda ssp.	Ind.	19	23	29	13	25	21	21	53	92	62
	ID%	3	6	7	5	12	10	93	130	119	76
	F%	38	44	58	33	58	71	17	56	75	112

Soil Compaction

Soil compaction decreases soil fertility and increases soil erosion. It is frequently caused by the use of heavy agricultural machines and by the reduced application of humus-producing substances such as organic manures. Investigations on the effects of soil compaction on soil protozoa were performed by Berger et al. and Coûteaux. The results of these studies demonstrate the inhibitory effects of even slight compaction on soil protozoa and the dramatic reduction of life in heavily compacted soils. The changes are very likely due to reduced pore space and lower moisture content. The ciliates appear to be more sensitive but less selective than the testaceans. Berger et al. performed their investigations with special compaction chambers that compacted 500 cm^{-3} of the upper soil layer of an alpine pasture on alpine pseudogley by 10, 30 and 50%. The chambers were exposed in the field in three parallel sets for three months at the same sites. The results showed a significant decrease in the abundance of the testaceans and the nematodes and a loss of the infrequent species with increasing compaction. Centropyxis aerophila var. sphagnicola was most abundant in the control and in the 10 and 30% compaction trials, whereas the smaller Trinema lineare dominated at 50% compaction, perhaps because of the reduced pore space. The number of species, the abundance of testate amoebae and nematodes, and the soil moisture content correlated negatively with the bulk density of the soil. Empty tests (dead individuals) increased with compaction, indicating a higher mortality and/or reduced decomposition. Active ciliates (direct counting method) were most abundant at 30% compaction. Coûteaux found the individual numbers of the ciliates significantly depressed in 4 g microcosms made of compacted forest humus.

Abundance, individual dominance, biomass dominance, and frequency of important testacean species in the litter layer (0–3 cm) of a declining, ameliorated spruce forest.

pH type	Parameter	Treatment					
		C	M	O	c	m	o
Acidtolerant species	ID%	47.5	28.2	23.2	44.0	28.8	27.5
Corythion dubium	Ind.	8724	6726	4857	13 575	7798	8100
	ID%	40.0	25.4	20.1	40.6	27.4	26.6
	BD%	7.5	6.7	5.3	14.2	9.5	12.6
	F%	97.4	100.0	100.0	97.2	100.0	97.2
Nebela spp.	Ind.	1628	734	756	1134	387	280
	ID%	7.5	2.8	3.1	3.4	1.4	0.9
	BD%	26.6	16.3	17.3	28.4	11.7	10.4
	F%	86.8	57.9	50.0	80.6	44.4	41.7
Circumneutral species	ID%	25.5	45.2	50.2	32.8	49.6	53.7
Cryptodifflugia oviformis	Ind.	935	3698	3097	2305	4587	4337
	ID%	4.3	14.0	12.8	6.9	16.1	14.2
	BD%	0.2	1.0	0.9	0.6	1.5	1.8
	F%	50.0	71.1	65.8	86.1	97.2	88.9
Trinema spp.	Ind.	4635	8263	9041	8663	9535	12 035
	ID%	21.2	31.2	37.4	25.9	33.5	39.5
	BD%	7.0	16.7	18.6	17.8	25.7	42.6
	F%	92.1	100.0	97.4	100.0	100.0	100.0

Movement of Protozoan Pathogens in Soil

Livestock wastes may contain pathogenic protozoa, especially sporozoans, and it is now recognized that, following the application of livestock waste to land, there is a potential for the transfer of these microorganisms to humans after contamination of plants, soil, and subsequently, water courses. Cryptosporidium parvum is of particular interest because as few as 10 infective oocysts may be required to cause infection, and its transmissive oocyst is, furthermore, resistant to current methods of drinking water treatment.

Mawdsley et al. investigated the potential for transfer of C. parvum through soil to land drains and, subsequently, water courses following the application of livestock waste to land. They used intact soil cores mounted on a special soil tilting table apparatus in the laboratory. Following irrigation over a 21-day period, Cryptosporidium parvum oocysts applied to the surface of soil cores (initial inoculum concentration 1×108 oocysts per core) were detected, albeit in low numbers, in the leachates from clay loam and silty.

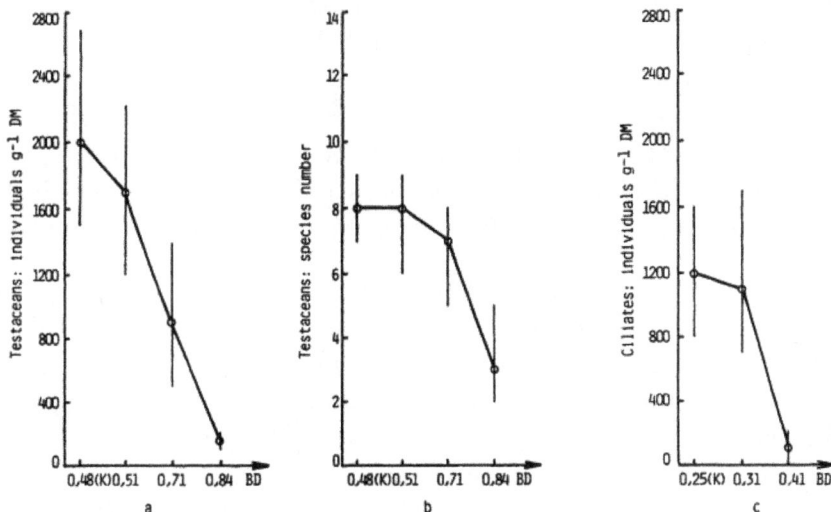

Effect of soil compaction on testate amoebae and ciliates. (a), (b): Field experiment with the very humic topsoil (0–5 cm) of an alpine pasture. Methods: randomized block design (bars: extreme values), 12 weeks compaction in chambers with 500 cm3, direct counts as described in Foissner; (c): Laboratory experiment with the humus layer of a mixed forest. Methods: 9 weeks compaction of 4 g humus in compaction chambers, direct counts and calculation with a most probable number method. BD: bulk density; DM: dry mass of soil; K: control.

loam soils, but not in the leachate from a loamy sand soil. Variations in leaching patterns were recorded between replicate cores. At the end of the study, soil cores were destructively sampled to establish the location of oocysts remaining within the soil. Distribution within cores was similar in all three soil types; the majority ($72.8 \pm 5.2\%$) of oocysts were found in the top 2 cm of soil, with numbers decreasing with increasing depth to $13.2 \pm 2.8\%$, $8.39 \pm 1.4\%$, and $5.36 \pm 1.4\%$ at depths of 10, 20, and 30 cm, respectively.

Soil Protozoan Bioassays

Although soil protozoa have frequently been used as test organisms definite test systems have only recently been developed. Both systems are very similar and measure the 24 h growth response of either Colpoda steinii or C. inflata, two very common and well-known soil ciliates.

These test systems have been applied with heavy metals in laboratory and field trials The results agree well and suggest that soil ciliates are at least as sensitive to environmental hazards as more commonly used test organisms (e.g., earthworms). There is thus a strong likelihood that protozoa can replace and/or supplement invertebrates and vertebrates in some assays. However, as with other organisms, results are highly dependent on procedural differences and other, not yet fully understood variables such as pH and organic compounds. Acute toxicity values for a given metal may vary by as much as almost two orders of magnitude, emphasizing the need for strict test standardization. Sensitivities of Colpoda spp. to heavy metals, as tested by Bowers et al. and Forge et al., tended to fall in the lower range for all metals, supporting the inclusion of a ciliate test in a 'battery of tests' approach to soil toxicity assessment. Furthermore, test results are obtained within 24 h, i.e., much faster than with any multicellular system, e.g., the standard 5-day seed germination test or the Eisenia fetida test endorsed by the EC, which requires 7–14 days to complete.

Effect of sewage sludge treatment on growth of Colpoda steinii in soil solution extracted from (a) Lee Valley and (b) Luddington unamended control soils, soils amended with uncontaminated sludge and soils amended with sewage sludge contaminated with either high or low amounts of Zn (16 000 or 8000 mg kg^{-1}, respectively), Ni (4000 or 2000 mg kg^{-1}, respectively) or Cu (8000 or 4000 mg kg^{-1}, respectively). Error bars = LSD$_{0.05}$. The test system of Forge et al. was used.

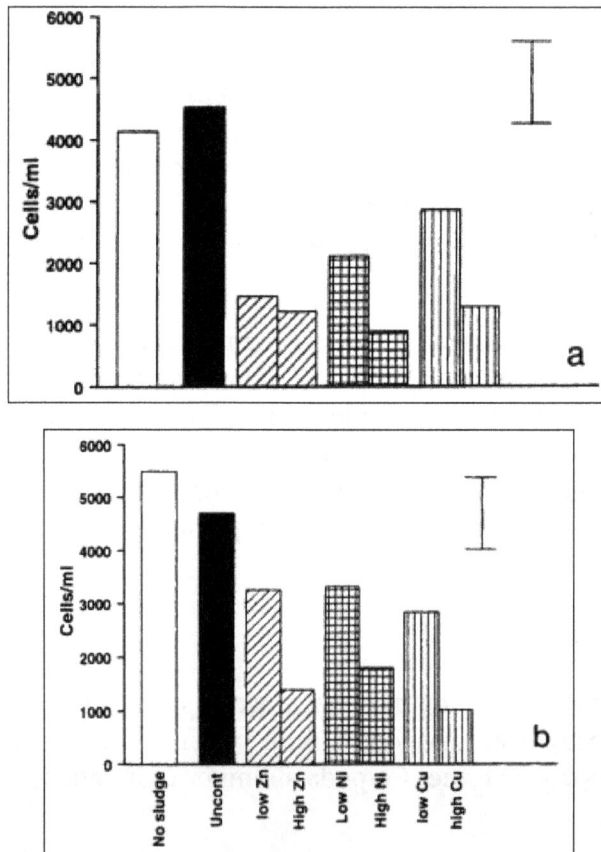

As an example, figure shows some of the results obtained by Campbell et al., who used the test system of Forge et al. Campbell et al. investigated the toxicity and bioavailability of heavy metals in soil solutions extracted from two long-term experiments that were amended with sewage sludge. The sludges were predominately contaminated with either Ni, Cu, or Zn. Growth of Colpoda steinii was strongly inhibited in all metal-amended soils compared to the equivalent unamended control soils. The EC50 values for relative growth were in order of decreasing toxicity Cu > Ni > Zn, which corroborates results by Madoni et al., who found Cu > Hg > Cd > Zn for activated sludge ciliates. Using the same assay to study the toxicity of metal sulphate solutions, Forge et al. found toxicity decreased in the order of Ni > Cu > Zn, which is similar to the ranking obtained by Pratt et al., who reported Cd > Cu > Zn.

References

- Biology, soils: nrcs.usda.gov, Retrieved 22 August, 2019

- Benefits-of-nematodes-in-healthy-soil-ecosystems: nwdistrict.ifas.ufl.edu, Retrieved 30 July, 2019

- Soil-nematodes-in-organic-farming-systems: extension.org, Retrieved 4 June, 2019

- Biology, health, soils: nrcs.usda.gov, Retrieved 15 April, 2019

Applications of Soil Microbiology

Soil microbiology is applied in a number of different areas, such as agriculture and for the production of biopesticides. It is also involved in studying the relationship between global climate change and microbes as well as the effect of heavy metal contamination upon the microorganisms in the soil. This chapter closely examines these important applications of soil microbiology to provide an extensive understanding of the subject.

Standardisation of Methods in Soil Microbiology

Microorganisms in soil ecosystems are ubiquitous, abundant, diverse and essential for many soil functions such as carbon and nitrogen cycling, plant productivity and climate regulation. Because of their importance, there is a large volume of past and contemporary researches that aims to understand the ecology of soil microbial communities, with thousands of articles devoted to this research field published annually. Numerous methods have been developed to estimate abundance, diversity and activity of soil microorganisms. Several such procedures are now successfully applied on a regular and on-going basis, perhaps most notably the chloroform fumigation-extraction technique for estimating microbial biomass, and DNA fingerprinting approaches for estimating the structure of microbial communities. Perversely, many of these methods become victims of their own success, and a plethora of laboratory- or even user-specific protocols, which contain minor to major modifications of the initially described methods, are now used worldwide. However, these differences between protocols are far from being inconsequential as they often include inherent bias, which hamper data comparison across studies, let alone laboratories. Indeed, variations in data obtained by different laboratories or using different protocols are commonly reported. A theoretically obvious, albeit practically challenging, solution is to define and use standardised methods. This is becoming all the more important because an exponentially increasing volume of data is now being generated, particularly with the advent of automated or high-throughput techniques, notably in relation to molecular biology. Such techniques offer exciting opportunities for better understanding soil microbial diversity, how it relates to soil functions, and more effective ways to manage terrestrial ecosystems to meet the challenges of sustainability. This grand challenge should be facilitated by ensuring comparable data, which is necessary in order that our knowledge of soil microbial communities can be effectively integrated.

The concept, and practice, of standardisation in soil microbiological assays can be applied at a range of levels, from the individual researcher/group (vital to ensure coherence within a body of experimentation), through institutional (assists integration and co-herence within institutional-level programmes), to national (e.g. British Standards and French National Organisation for Standardisation) and international [e.g. International Organisation for Standardisation, (ISO)]. Here we focus on the latter context, as this is arguably the most effective route to achieve the higher-level aims of standardisation. Moreover, science itself is an international collaborative effort and comparisons across studies need to be performed beyond country borders, not least because soils and the organisms they support operate entirely independently of such boundaries. Standards providing internationally agreed methods for assessing soil microorganisms have most-ly been developed by the International Organisation for Standardisation (ISO). How-ever, the number of ISO standardised methods is still scant in relation to the numerous methods that have been developed within the field of soil microbiology. In addition, the use of ISO methods in soil microbiology research articles, outside of ecotoxicology studies, is in our perception relatively rare. In this topic, we explain the importance of standardisation in soil microbiology, present an overview of the existing and forthcom-ing ISO standards, and discuss some technical and cultural hurdles. One aim is to stim-ulate debate in this field and to encourage a move toward the development and greater dissemination of internationally agreed standards in soil microbiology.

Dealing with the Natural Complexity and Diversity Soils are arguably the most complex systems on the planet, given the extraordinary diversity of their chemical and biological constituents, as well as the extreme struc-tural heterogeneity. There are also a wide range of soil types, with huge numbers of classes of soil recognised in taxonomic schemes both at global down to national scales, for example, some 748 Soil Series are recognised in the Soil Survey of England and Wales and thousands of types in the lower-order taxa of World Reference Base. The geo-spatial distribution of soils is also complex across virtually all size scales, which means that studies at almost any spatial scale involve a variety of soil types, which may confound the ready application of standard techniques. This diversity of constitution and basic characteristics severely challenges the ability to set standards in measuring soil properties and processes. This is particularly true for biological aspects of soil sys-tems, and in part accounts for the concomitant diversity in methodological variants. Even something as outwardly straightforward as determining soil organic carbon is confounded by the fact that soils can vary from essentially 0–100% organic matter, there is potential (and variable) interference from inorganic forms of carbon, and the same procedure is certainly not appropriate for soils at the two extremes. It is often then the case that no single method is universally appropriate and that variants within methods are needed to compensate for differences in properties that may occur if they are to be applicable to the gamut of soils. For example, measuring soil respiration by CO_2 emission is relatively straightforward if the pH of the soil is lower than 7.5, but in more alkaline soils, the partition coefficient of CO_2 between air and water starts to con-found the technique because proportionately more CO_2 will prevail in the pore water. The quality and quantity of organic matter and clay vary between soils that affects the nature and extent of potential absorption of biochemicals, notably nucleic acids, such that a range of devices to counter such effects need to be applied, contingent on the soil. These factors can be compensated for by variants in technique, and such variants can be duly standardised. In principle, such matters do not then preclude the setting of standards, but they certainly prevent the setting of simple standards. Furthermore, there is a significant issue that affects data comparability, as with complex protocols, there is an increased likelihood that different operators will determine different absolute values for measurements, because of accumulations of even subtle differences between each of the steps in such procedures.

Another factor arising from the need for sophisticated/adjusted/complex protocols is the ease with such protocols are agreed upon within the context of a standards setting framework, particularly an international one. This is because the optimal procedures are not necessarily readily defined and can become more a matter of best judgement. For example, it can be argued either way that the pH of the buffer medium in enzyme assays should be standardised to a particular pH, or the pH of the particular soil under scrutiny, but there are then supplementary issues of how to determine that pH. Another concern is at which temperature one should measure soil respiration? The same for a sub-arctic tundra soil as one from Namibia or a 'locally pertinent' temperature? And then what moisture content is optimal for respiration measurements and how should that be determined? Such questions are undoubtedly very important in defining standards but challenge the attainment of scientific consensus.

Current Standards in Soil Microbiology

Despite the inherent complexity and diversity of soils described earlier, some methods to study soil microorganisms have been standardised since 1997. Due to a strong concern regarding the degradation of soils in relation to local and diffuse contamination or loss of biodiversity, the existing standards were developed by the 'Soil quality'. Technical Committee ISO/TC 190 with a strong focus on assessing the effects of chemicals and pollution on the soil fauna and soil microorganisms. Methods for measuring soil microbial biomass using substrate-induced respiration and fumigation-extraction were the first ones to be standardised in the field of soil microbiology in the late nineties (ISO 14240,). Indeed, these methods based on pioneering work of Vance et al. were proposed to provide a sensitive indicator for measuring changes in the total quantity of soil microorganisms in response to environmental factors or anthropogenic disturbances. Most of the other existing ISO standards were developed for similar purposes and are therefore biased toward effective monitoring of the soil microbial community to meet extant policy requirements. This trend is particularly obvious for ISO 14238 'Determination of nitrogen mineralisation and nitrification in soils and the influence of chemical on these processes' and ISO 15473 'Testing for biodegradation of organic chemicals in soil'. Thus, ISO 14238 was designed to determine the effects of different concentrations of a chemical on the N-cycling processes using dose–response curves while ISO 15473 gives general guidelines for the selection and method of tests to determine the biological degradation of organic chemicals introduced into the soil either intentionally or accidentally.

Criteria related to applicability and effectiveness of standards for routine analyses such as high throughput analysis, cost, usability or data interpretation have up to now excluded molecular methods, such as terminal fragment length polymorphism for assessing microbial diversity, despite their widespread use in research. However, among the new ISO standards, the development of the ISO 11063 standard for soil DNA extraction is of special interest because it is the first step of all PCR-, hybridisation, and sequencing-based molecular analyses of the diversity and abundance of soil microbial communities. As a result, thousands of studies are performed yearly in environmental microbiology using soil DNA extraction methods. Due to this important business market, at least ten companies are commercialising soil DNA extraction kits, which add to the list of home-made protocol. This is despite it being well established that the apparent microbial diversity determined by any nucleic acid analysis procedure is contingent on the DNA extraction method. The ISO 11063 standard for soil DNA extraction is based on both chemical and physical approaches for extraction and lyses of the microbial cells as described by Petric et al. This ISO is timely since studies of soil microbial diversity based on soil DNA extraction are generating an exponential amount of sequence data, and large scale projects aiming at sequencing the soil metagenome are now launched. Knowledge of the identity and the quantity of each compound used in the ISO 11063 or any ISO protocol provides transparency and allow users a complete quality control, which is a major advantage over commercial kits. Thus, production batch effects can occur, and this has been observed for some commercial soil DNA extraction kits. A transparent protocol also avoids the risk of subsequent modifications of the kit reagents by companies or risks associated to the versatility of their business strategies such acquisition and merging, which are common activities for biotechnology industry.

While no nucleic acid-based method for assessing soil microbial diversity have yet been proposed for international standardisation, two lipid-based methods have recently became ISO standards. Phospholipid fatty acid (PLFA) and phospholipid ether lipids (PLEL) analyses are rapid and inexpensive methods for providing a quantitative measure of the viable soil biomass and complex microbial community profiles. They offer the advantage of targeting the entire microbial community, thus allowing calculation of the fungal/bacteria ratio using markers PLFA specific of these domains. Since the late 1990s, several comprehensive reviews discussing the strengths and weaknesses of the use of lipid fatty acids for assessing microbial biomass and community structure in soil have been published. Unfortunately, while some of the ISO standards described in table have been published more than 10 years ago, their use by the scientific community is still very limited. Thus, the ISO has no power to enforce the implementation of the standards it develops and therefore adoption of the ISO standard is still mainly voluntary.

Directions for Future Standards

The standardisation effort is uneven between methods addressing the abundance, the diversity and the activity of the soil microbial community. Indeed, while there are already three ISO standards for quantifying soil microbial biomass, a new work item proposing a standard to estimate the abundance of the soil bacterial community by 16S rRNA gene targeted quantitative PCR (qPCR) was recently adopted by the Soil quality ISO technical committee. The recent developments of qPCR analyses also allow the quantification of the abundances of specific functional or taxonomical microbial groups, which may represent useful bioindicators. With the use of appropriate blanks, internal and surrogate standards, qPCR is a reliable method having the advantage to offer high throughput and cost-effective analyses.

For a better understanding of soil microbial activity, or more generally of soil functioning, several methods for quantifying potential enzyme activity have been developed. Even though these methods providing an insight of the size of the enzyme pool have some limits, they are commonly used as microbiological indicators of soil quality and should therefore be standardised for comparison of microbial activities both between soils and laboratories. For example, because of their environmental and agronomical importance, microorganisms involved in N-cycling are of key interest. In addition, they are popular models in soil microbial ecology for relating microbial diversity and soil functioning. However, only measurement of potential nitrification has been internationally standardised up to now, while methods for monitoring other N-processes such as nitrogen fixation and denitrification also necessitate standardisation. For example, the original protocol for estimating potential denitrification has been modified in many ways. In this assay, to measure the activity of the pool of denitrification enzymes in the soil at the time of sampling, soil slurries are incubated in the laboratory in non-limiting denitrification conditions (without oxygen, addition of nitrate and carbon, and of chloramphenicol to avoid de novo synthesis) so that only the amount of enzyme is rate-limiting. Changes in the original protocol include excluding the chloramphenicol, which can decrease the activity of synthesized enzymes, addition of different carbon types and amount (glucose, acetate, glumatic acid, etc) and incubation of the soil slurries in various conditions. Similarly, determination of the nitrogenase activity using the acetylene reduction technique is subjected to various modifications of the protocol resulting, for example, in variants of the acetylene concentration (0.03–0.1 v/v). In contrast to other methods, most modifications of these methods are not soil-specific and both potential denitrification and nitrogen-fixation assays could readily be standardised in future.

Finally, regarding methods to monitor the diversity and the structure of the soil microbial community, the adoption of the ISO 29843 for PLFA and PLEL analyses opens the path for other standards. While it is too early to propose any standardisation of the new high-throughput sequencing technologies (e.g. 454 pyrosequencing), other powerful approaches such as those based on taxonomic and functional microarrays meet the criteria to become standards. Of course these perspectives for the development of future standards in soil microbiology are not exhaustive, and we encourage soil microbiologists to expand it by proposing other popular methods for standardisation.

Soil Microorganisms and Global Climate Change

Microorganisms found in the soil are vital to many of the ecological processes that sustain life such as nutrient cycling, decay of plant matter, consumption and production of trace gases, and transformation of metals. Although climate change studies often focus on life at the macroscopic scale, microbial processes can significantly shape the effects that global climate change has on terrestrial ecosystems. According to the International Panel on Climate Change (IPCC) report, warming of the climate system is occurring at unprecedented rates and an increase in anthropogenic greenhouse gas concentrations is responsible for most of this warming. Soil microorganisms contribute significantly to the production and consumption of greenhouse gases, including carbon dioxide (CO_2), methane (CH_4), nitrous oxide (N_2O), and nitric oxide (NO), and human activities such as waste disposal and agriculture have stimulated the production of greenhouse gases by microbes. As concentrations of these gases continue to rise, soil microbes may have various feedback responses that accelerate or slow down global warming, but the extent of these effects are unknown. Understanding the role soil microbes have as both contributors to and reactive components of climate change can help us determine whether they can be used to curb emissions or if they will push us even faster towards climatic disaster.

Microbial Contributions to Greenhouse Gas Emissions

Soil microorganisms are a major component of biogeochemical nutrient cycling and global fluxes of CO_2, CH_4, and N. Global soils are estimated to contain twice as much carbon as the atmosphere, making them one of the largest sinks for atmospheric CO_2 and organic carbon. Much of this carbon is stored in wetlands, peatlands, and permafrost, where microbial decomposition of carbon is limited. The amount of carbon stored in the soil is dependent on the balance between carbon inputs from leaf litter and root detritus and carbon outputs from microbial respiration underground. Soil respiration refers to the overall process by which bacteria and fungi in the soil decompose carbon fixed by plants and other photosynthetic organisms and release it into the atmosphere in the form of CO_2. This process accounts for 25% of naturally emitted CO_2, which is the most abundant greenhouse gas in the atmosphere and the target of many climate change mitigation efforts. Small changes in decomposition rates could not only affect CO_2 emissions in the atmosphere, but may also result in greater changes to the amount of carbon stored in the soil over decades.

Methane is another important greenhouse gas and is 25 times more effective than CO_2 at trapping heat radiated from the Earth. Microbial methanogenesis is responsible for both natural and human-induced CH_4 emissions since methanogenic archaea reduce carbon into methane in anaerobic, carbon-rich environments such as ruminant livestock, rice paddies, landfills, and wetlands. Not all of the methane produced ends up in the atmosphere however, due to methanotrophic bacteria, which oxidize methane into CO_2 in the presence of oxygen. When methanogens in the soil produce methane faster than can be used by methanotrophs in higher up oxic soil layers, methane escapes into the atmosphere. Methanotrophs are therefore important regulators of methane fluxes in the atmosphere, but their slow growth rate and firm attachment to soil particles makes them difficult to isolate. Further exploration of these methanotrophs' nature could potentially help reduce methane emissions if they can be added to the topsoil of landfills, for example, and capture some of the methane that would normally be released into the atmosphere.

Not unlike their role in the carbon cycle, soil microorganisms mediate the nitrogen cycle, making nitrogen available for living organisms before returning it back to the atmosphere. In the process of nitrification (during which ammonia is oxidized to nitrate), microbes release NO and N_2O, two critical greenhouse gases, into the atmosphere as intermediates. Evidence suggests that humans are stimulating the production of these greenhouse gases from the application of nitrogen-containing fertilizers. For example, Nitrosomonas eutropha is a nitrifying proteobacteria found in strongly eutrophic environments due to its high tolerance for elevated ammonia concentrations. N-fertilizers increase ammonia concentrations, causing N. eutropha to release more NO and N_2O in the process of oxidizing ammonium ions. Since NO is necessary for this reaction to occur, its increased emissions cause the cycle to repeat, thereby further contributing to NO and N_2O concentrations in the atmosphere.

Microbial Responses to Global Climate Change

Microbial processes are often dependent on environmental factors such as temperature, moisture, enzyme activity, and nutrient availability, all of which are likely to be affected by climate change. These changes may have greater implications for crucial ecological processes such as nutrient cycling, which rely on microbial activity. For example, soil respiration is dependent on soil temperature and moisture and may increase or decrease as a result of changes in precipitation and increased atmospheric temperatures. Due to its importance in the global carbon cycle, changes in soil respiration may have significant feedback effects on climate change and severely alter aboveground communities.

Microbial Response to Increased Temperatures

One of the major uncertainties in climate change predictions is the response of soil respiration to increased atmospheric temperatures. Several studies show that increased temperatures accelerate rates of microbial decomposition, thereby increasing CO_2 emitted by soil respiration and producing a positive feedback to global warming. Under this scenario, global warming would cause large amounts of carbon in terrestrial soils to be lost to the atmosphere, potentially making them a greater carbon source than sink However, further studies suggest that this increase in respiration may not persist as temperatures continue to rise. In a 10-year soil warming experiment, Melillo et al. show a 28% increase in CO_2 flux in the first 6 years of warming when compared to the control soils, followed by considerable decreases in CO_2 released in subsequent years, and no significant response to warming in the final year of the experiment. The exact microbial processes that cause this decreased long-term response to heated conditions have not been proven, but several explanations have been proposed. First, it is possible that increased temperatures cause microbes to undergo physiological changes that result in reduced carbonuse efficiency. Soil microbes may also acclimate to higher soil temperatures by adapting their metabolism and eventually return to normal decomposition rates. Lastly, it can be interpreted as an aboveground effect if changes in growing season lengths as a result of climate change affect primary productivity, and thus carbon inputs to the soil.

The effects of increased global temperatures on soils is especially alarming when considering the effects it has already begun to have on one of the most important terrestrial carbon sinks: permafrost. Permafrost is permanently frozen soil that stores significant amounts of carbon and organic matter in its frozen layers. As permafrost thaws, the stored carbon and organic nutrients become available for microbial decomposition, which in turn releases CO_2 into the atmosphere and causes a positive feedback to warming. One estimate suggests that 25% of permafrost could thaw by 2100 as a result of global warming, making about 100 Pg of carbon available for microbial decomposition. This could have significant effects on the global carbon flux and may accelerate the predicted impacts of climate change. Moreover, the flooding of thawed permafrost areas creates anaerobic conditions favorable for decomposition by methanogenesis. Although anaerobic processes are likely to proceed more slowly, the release of CH_4 into the atmosphere may result in an even stronger positive feedback to climate change.

Microbial response to increased CO_2

Atmospheric CO_2 levels are increasing at a rate of 0.4% per year and are predicted to double by 2100 largely as a result of human activities such as fossil fuel combustion and land-use changes. Increased CO_2 concentrations in the atmosphere are thought to be mitigated in part by the ability of terrestrial forests to sequester large amounts of CO_2. To test this, an international team of scientists grew a variety of trees for several years under elevated CO_2 concentrations. They found that high CO_2 concentrations accelerated average growth rate of plants, thereby allowing them to sequester more CO_2. However, this growth was coupled with an increase in soil respiration due to the increase in nutrients available for decomposition by releasing more CO_2 into the atmosphere. This suggests that forests may sequester less carbon than predicted in response to increased CO_2 concentrations, however more research is needed to investigate this hypothesis.

According to the IPCC (2007) report, climate change will alter patterns of infectious disease outbreaks in humans and animals. Soil pathogens are no exception: case studies support the claim that climate change is already changing patterns of infectious diseases caused by soil pathogens. For example, over the last 20 years, 67% of the 110 species of harlequin frogs (Atelopus) native to tropical regions in Latin America have gone extinct from chytridiomycosisthe, a lethal disease spread by the pathogenic chrytid fungus (Batrachochytrium dendrobatidis). Research suggests that mid- to high-elevations provide ideal temperatures for B. dendrobatidis. However, as global warming progresses, B. dendrobatidis is able to expand its range due to increasing moisture and warmer temperatures at higher elevations. This expansion exposes more amphibian communities in previously unaffected or minimally affected areas, specifically at higher elevations, to chytridiomycosisthe. As seen in the case of Atelopus harlequin frogs, the spread of soil pathogens due to climatic changes can significantly affect life at the macro scale and ultimately lead to species extinction.

Global soils already hold three times as much carbon as exists in the atmosphere, and there's room for much more. According to a recent study in Nature, enhanced carbon storage in the world's farmland soils could reduce greenhouse gas concentrations by between 50 and 80 percent.

To realize this stunning potential, farmers would need to adopt certain game-changing farming practices that restore depleted soils, largely through spurring the activity of the soil microbiome, a web of microscopic life that includes fungi, nitrogen-fixing bacteria and trillions of other bacteria that promote plant growth. Like the microbes that live in and on our bodies, helping us with everything from nutrition to immune responses, soil microbes are allies. They can help us deal with many of the climate challenges facing agriculture.

Indeed, we are just beginning to understand how to harness the potential of soil microbes. Research has shown they can help restore degraded soils, including land in Mexico's southern Sonoran desert. This capacity gives soil microbes the potential of revolutionize agriculture. Healthier soils produce higher crop yields, hold water more effectively, sequester more carbon and allow for increased agricultural productivity on existing land.

Secondly, soil microbes can help plants tolerate hot temperatures and drought brought about by climate change. Recent research has shown that soil microbes can help plants like wheat, rice, pepper and maize to withstand drought. Plants treated with soil microbes have a deeper root system and their shoots grow more quickly. Consequently, under drought stress, plants inoculated with microbes can more effectively take up water from drying soil and maintain near-normal shoot growth rates resulting in increased crop productivity.

Thirdly, soil microbes increase plant defenses against insect pests whose populations are expected to increase due to the changing climate. In India, researchers have shown that soil microbes applied directly to seeds helped plants combat the rice leaf-folder insect, an important rice pest in China, Japan, the Philippines and Vietnam. In another study, treatment of cotton plants with soil microbes helped them fight off beet army-worm, by killing its larvae.

What's more, soil microbes can improve overall plant growth. This is especially important to the world's 500 million smallholder farmer families, many of whom live in Africa and produce one-quarter of the average global yield of cereal crops. Increased productivity and income would power a virtuous cycle, enabling poor farmers to invest even more in the sustainability and productivity of their farms. The use of soil microbes to improve soil health and mitigate climate change would be invaluable in parts of the developing world hardest hit by drought and rising temperatures.

The challenge is to develop products that work for these smallholder farmers. To date, the handful of soil-microbial products that have reached the market are being manu-factured by big companies like BASF, Syngenta and Monsanto. After spending millions of dollars to make these products, which were created for the world major cash crops such as soy and corn, they are unlikely to come at a cost that many small-scale farmers can afford.

Furthermore, to ensure that smallholder farmers benefit from new biological products, research is needed to map out the diversity of microbes in different crops and climates. Once identified, industry needs to develop cheaper methods to grow the microbes on a scale that would be available to millions of farmers.

We can't expect private sector companies alone to undertake the research and prod-uct development needed to serve poor farmers across Africa and Asia. Rather, private companies, public research institutions, governments and other partners need to work together to deliver soil microbial products useful to small-scale farmers in developing countries.

Effects of Heavy Metal Contamination upon Soil Microbes

Heavy metal contaminants in the environment are eventually deposited in soils in some form of a low solubility compound, such as pyrite or sorbed on surface-reactive phases, such as Fe and Mn oxides. While this phenomenon immobilizes the contaminants, thus limiting their effects upon biota and human health, it also places metal ions in an inti-mate contact with soil microbial community.

A number of studies were completed on the role of soil/sediment microbial commu-nity in heavy metal scavenging and release in recent decades. Those studies highlight the importance of metal-microbe interactions and microbial processes from sorption and desorption of metals on the cell surfaces to reduction/oxidation phenomena in determining the fate of heavy metals in the environment. The opposite side of the interaction, namely, how heavy metals at lower concentrations impact composition of a microbial community and functioning of specific pathways within that community is substantially less understood, especially in the circumneutral environments where metal levels are low and metals are not exceedingly mobile. Our knowledge of metal effects at lower, more environmentally relevant conditions, circumneutral pH, and response of metagenome to heavy metal challenge under these conditions is still limited to a handful of studies. In contrast, highly acidic, high-metal conditions, such as those occurring in mine tailings have been thoroughly investigated.

Although some heavy metals are required for life's physiological processes (e.g., components of metalloenzymes), their excessive accumulation in living organisms is always detrimental. Generally, toxic metals cause enzyme inactivation, damage cells by acting as antimetabolites or form precipitates or chelates with essential metabolites. The impact of some toxic metals on human health has been reported by Forstner. Human diseases have resulted from consumption of cadmium-contaminated foods. Likewise, Pb exposure can cause seizures, mental retardation, and behavioral disorders. The threat that heavy metals pose to human and animal health is aggravated by their low environmental mobility, even under high precipitations, and their long-term persistence in the environment. For instance, Pb, one of the more persistent metals, was estimated to have a soil retention time of 150 to 5000 years. Also, the average biological half-life of Cd, another accumulation poison similar to lead, has been estimated to be about 18 years. As a result of low environmental mobility of those metals, a single contamination episode could set a stage for a long-term exposure of microbial community to metal, necessitating a long-term monitoring effort to assess metal effects.

Studies have shown that long-term heavy metal contamination of soils has harmful effects on soil microbial activity, especially microbial respiration. Aside from long-term metal-mediated changes in soil enzyme activities, many reports have shown large reductions in microbial activity due to short-term exposure to toxic metals. Bacterial activity, measured by thymidine incorporation technique, had been shown to be very sensitive to metal pollution both under laboratory and field conditions. Moreover, habitats that have high levels of metal contamination show lower numbers of microbes than uncontaminated habitats.

Enzyme activity is a soil property that is chemical in nature but has a direct biological origin. This activity arises from the presence of many types of enzymes that are present in the soil, and within soil microorganisms. From an assortment of enzymes present and active in soil, phosphatases are interesting groups of enzymes that catalyze the hydrolysis of phosphate from organic monoester linkages. Phosphates released from such phosphatase action are very important to the plants and microorganisms that depend on soil for their phosphorus requirements.

It is well established that toxic effects of heavy metals are highly selective in the higher organisms. Specific organ targeting was shown for mercury and silver in invertebrates. Indications of specific inhibitory action of heavy metals have been produced in microbes as well. Such selective targeting of specific enzymatic systems and pathways suggests that certain members of the microbial community would be more sensitive to heavy metal exposure than others, depending on the sensitivity of their critical metabolic pathways. Thus, while toxicity of heavy metals to microbes is a well established phenomenon, the effects of those metals upon specific enzymatic systems at lower ("sub-acute") concentrations are not well known.

Denitrification is a natural microbial process converting nitrate to dinitrogen gas during anaerobic respiration. Such reduction occurs sequentially, with nitrate converted to nitrite, nitric oxide, nitrous oxide and, finally, nitrogen gas. A number of enzyme classes, mostly located in the periplasmic space, are involved in denitrification, with a number of corresponding genes that can be used as genetic markers for presence and expression of such enzymes in the soil metagenome. Based on their function, those enzymes can be broadly classified into nitrate reductases (converting nitrate to nitrite, narG and napA genetic markers correspond to those enzymes), nitrite reductases (nitrite to nitric oxide, nirK and nirS markers), nitric oxide reductases (nitric oxide to nitrous oxide, norB marker) and nitrous oxide reductases (nitrous oxide to dinitrogen gas, nosZ marker). The group of organisms capable of denitrification is not monophyletic; extensive horizontal gene transfer occurred early in the evolution of the pathway, resulting in nearly identical gene and enzyme forms in very distantly related organisms. As a result, 16S rRNA, which serves as a standard marker to assess microbial community diversity, is essentially useless for studying denitrifying bacteria.

Effects of heavy metals upon the community of microbes responsible for nitrogen cycle are still largely unknown. Stephen et al. analyzed the diversity changes of ammonia-oxidizing Betaproteobacteria exposed to metals at substantial (over 50 ppm) concentrations by studying amoA ammonium monooxygenase marker. Cloning and sequencing approach taken by those authors led to identification of large phylogenetic clusters of organisms responsible for ammonia oxidation in the presence of metal. However, no information was gained on the denitrifying community in this or other studies. Likewise, analyses of phospholipid-linked fatty acids commonly used in microbial community analysis provides no insight into effects upon phylogenetically incoherent physiological groups of microbes.

There are two basic strategies for a microbe to function in metal-contaminated environment. One, a system of transmembrane metal pumps has evolved in a number of bacteria, for example, system encoded by sil and mer operons, conferring resistance to silver and mercury, respectively. Those pumps scavenge metals on the inside of the cell membrane and remove them from the cell, thus protecting the internal cell structures from toxic metal effects.

Heavy metal resistance by means of metal ion scavenging and removal, such as provided by the classic mer or sil operons may not necessarily provide a sufficient degree of protection in the denitrifying organisms. As denitrification-related enzymes are generally located within the cell membrane or periplasmic space, expelling heavy metal ions out of the cell would place them in the immediate contact with denitrification-related enzymes, thus limiting utility of such a resistance strategy.

The fact that denitrification enzymes are located on or near the outer cell surfaces further increases the vulnerability of the entire denitrification pathway to chemical disruption. Recent work has suggested a direct effect of heavy metals upon extracellular enzyme activities. Combined with the fact that scavenging/pumping systems are unlikely to protect the denitrification pathway from heavy metal effects (and may, in fact exacerbate the situation), it is expected that denitrification pathway would be uniquely sensitive to heavy metals. The notion of selective inhibition of denitrification steps by heavy metals has been supported by work of Holtan-Harwig et al., suggesting the potential for production of undesirable byproducts, such as nitrous oxide.

The second mechanism of microbial resistance to metals is evolution of enzyme forms resistant to metals. This resistance pathway is expected to be the predominant in the denitrifying bacteria, due to inability to use metal pumps for the reasons described above. The metal-resistant forms of enzymes present in metal-stressed denitrifying community are expected to be readily identifiable by their gene sequence and therefore their genetic signature.

Disruption of denitrification by heavy metals could lead to a number of undesirable consequences, influencing the human health at both global and local levels. Suppressed denitrification in the soil could lead to enhanced nitrogen retention and flushing, resulting in nonpoint nutrient pollution in waterways receiving overland or subsurface flow from impacted locations. Nutrient pollution, in turn, leads to eutrophication and massive algal blooms, including those of toxic algae and cyanobacteria (e.g., Microcystis), affecting human populations relying on surface waters for municipal, recreational or agricultural needs. Specific inhibition of nitrous oxide reductase by metal has been observed recently, resulting in incomplete denitrification leading to emission of nitrous (and possibly nitric) oxides. As nitrous oxide is a potent greenhouse gas that also damages ozone layer, denitrification disruption via metal contamination could act as a link between local metal contamination and global climate change phenomena.

In the present study, we used molecular tools to assess the effects of low (1 ppm) and high (500 to 2000 ppm) of lead upon the denitrifying microbial community, as well as to assess applicability of the classical intermediate disturbance concept to microbial communities.

Permissions

All chapters in this book are published with permission under the Creative Commons Attribution Share Alike License or equivalent. Every chapter published in this book has been scrutinized by our experts. Their significance has been extensively debated. The topics covered herein carry significant information for a comprehensive understanding. They may even be implemented as practical applications or may be referred to as a beginning point for further studies.

We would like to thank the editorial team for lending their expertise to make the book truly unique. They have played a crucial role in the development of this book. Without their invaluable contributions this book wouldn't have been possible. They have made vital efforts to compile up to date information on the varied aspects of this subject to make this book a valuable addition to the collection of many professionals and students.

This book was conceptualized with the vision of imparting up-to-date and integrated information in this field. To ensure the same, a matchless editorial board was set up. Every individual on the board went through rigorous rounds of assessment to prove their worth. After which they invested a large part of their time researching and compiling the most relevant data for our readers.

The editorial board has been involved in producing this book since its inception. They have spent rigorous hours researching and exploring the diverse topics which have resulted in the successful publishing of this book. They have passed on their knowledge of decades through this book. To expedite this challenging task, the publisher supported the team at every step. A small team of assistant editors was also appointed to further simplify the editing procedure and attain best results for the readers.

Apart from the editorial board, the designing team has also invested a significant amount of their time in understanding the subject and creating the most relevant covers. They scrutinized every image to scout for the most suitable representation of the subject and create an appropriate cover for the book.

The publishing team has been an ardent support to the editorial, designing and production team. Their endless efforts to recruit the best for this project, has resulted in the accomplishment of this book. They are a veteran in the field of academics and their pool of knowledge is as vast as their experience in printing. Their expertise and guidance has proved useful at every step. Their uncompromising quality standards have made this book an exceptional effort. Their encouragement from time to time has been an inspiration for everyone.

The publisher and the editorial board hope that this book will prove to be a valuable piece of knowledge for students, practitioners and scholars across the globe.

Index

A

Acidic Soils, 12, 144-145, 160

Actinobacteria, 27, 29, 32, 74, 109, 114-120, 159

Actinomycetes, 4-5, 24, 45-47, 49-51, 54, 58, 85-86, 93, 109-110, 113, 115, 129, 139-142, 159, 179, 184

Aerobic Bacteria, 89, 110

B

Bacillus Subtilis, 178

Bacillus Thuringiensis, 47, 55

Biogeochemical Process, 46, 63

Biological Nitrogen Fixation, 46-47, 54, 128, 153, 155

Brownian Movement, 17-18, 183

C

Cation Exchange Capacity, 20, 54, 178

Ciliate, 109, 220, 223, 225-227, 229, 232

Clay Soil, 109, 112-113, 180, 183

Cyanobacteria, 32, 53-55, 111, 153, 246

D

Decomposer, 217

Denitrification, 23, 26, 46, 90, 112, 153, 238, 245-246

Deuteromycota, 161, 164

E

Ectomycorrhizae, 164-165, 167-171

F

Fabaceae, 128-129

Fungal Hyphae, 41-42, 54, 140, 146, 151, 170, 172, 174, 201, 216

Fungal-feeding Nematodes, 200-202, 206-207

G

Gram Staining, 128, 158

Grassland Soil, 27, 31, 40, 106

H

Heliothis, 47

L

Loam, 3, 7, 9, 115, 183, 186-187, 195, 231

M

Microbial Activity, 1, 4, 24, 26-27, 35, 37-38, 65, 87, 150, 213, 238, 240, 244

Microbial Biomass, 23-24, 52, 59-62, 96, 102-105, 107, 192, 207, 234, 236-238

Microbial Biomass Nitrogen, 60

Mineralization, 24, 38-39, 41, 46, 54-55, 62-63, 72, 76-77, 83-84, 91, 102, 104, 203, 206-208, 216-217

Mycorrhizal Fungi, 41-42, 64, 80, 89, 101, 148, 152-153, 161-162, 164, 171, 173, 200

Myrothecium, 85

N

Nematodes, 4-5, 40, 42-43, 45-46, 48, 51-55, 86, 90, 98, 101, 112 114, 120, 165, 199-210, 213-215, 220, 225, 228-230, 233

Nitrification, 20, 23, 111-112, 236, 238, 240

O

Organic Matter Decomposition, 4-5, 37, 44-45, 50, 86

P

Penicillium, 47, 85, 92, 101

Periplasmic Space, 245-246

Permafrost Soils, 30-32

Permeability of Soil, 5, 10

Plant-microbe Interaction, 76

Polymerase Chain Reactions, 48

Pore Space, 9-10, 35-36, 56, 203, 229-230

Porosity of Soil, 5, 9-10, 92

Predatory Nematodes, 200-202, 205-209

Pseudomonas, 32, 46-48, 50, 64, 66-67, 77, 85, 92, 99-100, 102, 106-108, 119, 184-185

R

Rhizobium, 45, 47, 50, 81, 111-112, 121-124, 126-129, 134-135, 159, 176-177, 186-187

Rhizoctonia, 45, 53, 85, 119, 121, 162

Rhizosphere, 27, 34-35, 37, 39, 50 66-68, 72, 75, 77, 79-80, 94, 102-107, 109, 113, 115, 126, 138, 145, 147-150, 161, 173, 176, 216

S

Sandy Loam, 3, 7, 115, 183, 186

Saprophytic Fungi, 162, 164

Silt, 2, 6-8, 12, 92, 109, 113, 195

Soil Biota, 40-43

Soil Ciliate, 220

Soil Compaction, 42-43, 229, 231

Soil Density, 5, 9

Soil Disturbance, 62, 85, 113

Soil Ecosystem, 45, 95, 97, 162, 205

Soil Enzyme, 23, 106, 244

Soil Fertility, 20, 23-24, 154, 206, 218, 225, 229

Soil Food Web, 32-33, 40, 42, 95, 114, 175, 199, 202, 204-205, 209

Soil Health, 75, 96, 100, 223, 243

Soil Metagenomics, 25, 28-31

Soil Microbiome, 75, 242

Soil Organic Matter, 4, 15, 24, 37-38, 59, 62, 86-87, 90, 92-93, 96-97, 102, 150, 160, 173, 179, 203, 207

Soil Organosulphur Pool, 104

Soil Plasticity, 5, 11

Soil Profile, 26, 55, 59, 88, 144, 160

Soil Reaction, 20, 23

Soil Structure, 4, 8, 38, 44-45, 54-56, 88, 111, 113, 162, 173, 177, 212

Soil Temperature, 11, 182, 240

Soil Texture, 1, 3, 7-8, 92, 180

Springtails, 40, 53, 86, 228

Sulphur Cycling, 102-103, 105-106

Symbiotic Bacteria, 64

T

Testate Amoebae, 214, 216, 218-219, 221-222, 225, 227, 230-231

Thiobacillus, 47, 50

Trichoderma, 45-47, 85, 93-94, 99-101, 163

V

Verticillium, 53, 85, 121

W

Weathering, 1, 8-9, 12

X

Xanlhomonas, 45

Z

Zygomycota, 161, 164

www.ingramcontent.com/pod-product-compliance
Lightning Source LLC
Chambersburg PA
CBHW061936190326
41458CB00009B/2753